WORLDSCAPE
世界园林

No.3 2013

中国林业出版社

本辑主题：住宅景观
THEME: HOUSING LANDSCAPE

运河岸上的院子－泰禾红御西区 6 栋大宅园林景观工程
COURTYARD ON THE GRAND CANAL THE LANDSCAPE ENGINEERING OF WEST 6 MANSION HOUSES IN THAIHOT HONGYU

对话大师：亨利·巴瓦
MASTER DIALOGUE: HENRI BAVA

垂直绿化工法
VERTICAL GARDENING CONSTRUCTION TECHNIQUES

"园冶杯"风景园林

The 2014 "Yuan Ye Award" International

"园冶杯"风景园林（毕业作品、论文）国际竞赛是由中国建设教育协会和中国花卉园艺与园林绿化行业协会主办，中国风景园林网和《世界园林》杂志社承办，在风景园林院校毕业生中开展的一项评选活动。

时间安排
报名截止日期：2014年4月30日
资料提交截止日期：2014年6月10日

参赛资格
应届毕业生（本科、硕士、博士）

参赛范围
风景园林及相关专业的毕业作品、论文均可报名参赛

竞赛分组
竞赛设置四类：风景园林设计作品类、风景园林规划作品类、园林规划设计论文类、园林植物研究论文类。每类设置两组：本科组和硕博组

Time schedule
The application deadline is April 30,2014;Submission deadline is June 10, 2014

Qualification
This year's graduates (Bachelor,Master,Doctor)

Competition content:
Landscape architecture and the related specialized graduation work, or paper

Types
There are four types:The design works of landscape architecture,the planning works of landscape architecture, the design and planning papers of landscape architecture,and the papers of garden plants. Every type is set into two groups: the undergraduate group,the master and doctor group.

毕业作品、论文）国际竞赛
cape Architecture Graduation Project / Thesis Competition

The 2014 "Yuan Ye Award" International Landscape Architecture Graduation Project is hosted by China Construction Education Association, Chinese Flowers Gardening and Landscaping Industry Association, and is undertaken by China Landscape Architecture network and *Worldscape*. It is a competition among graduates in landscape architecture schools.

地址：北京市海淀区三里河路17号甘家口大厦1510
邮编：100037　电话：(86) 010-88364851
传真：(86) 010-88365357
学生咨询邮箱：yyb@chla.com.cn
院校咨询邮箱：messagefj@126.com
官方网站：http://www.chla.com.cn

Address: 1510#, Ganjiakou Building, No.17, Sanli River Road, Haidian District, Beijing.
Post Code: 100037　**Tel:** 010-88364851
Fax: 010-88365357
Student Advisory E-mail: yyb@chla.com.cn
College Advisory E-mail: messagefj@126.com
Website: http://www.chla.com.cn

世界园林
WORLDSCAPE

主办单位	亚洲园林协会	Host Organizations
	国际绿色建筑与住宅景观协会	Asian Landscape Association
	中国花卉园艺与园林绿化行业协会	International Association of Green Architecture and Residential Landscape China
		Hortiflora and Landscaping Industry Association

（按姓氏字母顺序排名）

总 编	王小璘（台湾）	Editor-in-Chief
		Xiaolin Wang（Taiwan）
副总编	包满珠 李 敏 沈守云 王 浩 周 进 朱育帆	Deputy Editors
		Manzhu Bao Min Li Shouyun Shen Hao Wang Jin Zhou Yufan Zhu
顾问编委	凌德麟（台湾） 罗哲文	Consultants
		Delin Ling（Taiwan） Zhewen Luo
编委会		Editorial Board
常务编委	Jack Ahern（美国） 曹南燕 陈蓁蓁 高翅 Christophe Girot（瑞士）	Managing Editors
	Karen Hanna（美国） 何友锋（台湾） 贾建中 况 平 Eckart Lange（英国）	Jack Ahern（USA） Nanyan Cao Zhenzhen Chen Chi Gao Christophe Girot（Switzerland）
	李如生 李 雄 李炜民 Patrick Miller（美国） 欧圣荣（台湾） 强 健	Karen Hanna（USA） Youfeng He（Taiwan） Jianzhong Jia Ping Kuang Eckart Lange（England）
	Phillippe Schmidt（德国） Alan Tate（加拿大） Henri Bava（法国） 王庚飞	Rusheng Li Xiong Li Weimin Li Patrick Miller（USA） Shengrong Ou（Taiwan） Jian Qiang
	王良桂 王向荣 谢顺佳（香港） 杨重信（台湾） 喻肇青（台湾） 章俊华 张 浪	Phillippe Schmidt（Germany） Alan Tate（Canada） Henri Bava（France） Geng fei Wang Lianggui Wang
	赵泰东（韩国） 周 进 朱建宁 朱育帆	Xiangrong Wang Shunjia Xie（HongKong） Chongxin Yang（Taiwan） Zhaoqing Yu（Taiwan）
		Junhua Zhang Lang Zhang Taidong Zhao（Korea） Jin Zhou Jianning Zhu Yufan Zhu
编 委	白祖华 陈其兵 成玉宁 杜春兰 方智芳（台湾） 黄 哲 简仔贞（台湾）	Senior Editors
	金晓玲 李春风（马来西亚）李建伟 李满良 林开泰（台湾） 刘纯青 刘庭风	Zuhua Bai Qibing Chen Yu-ning Cheng Chunlan Du Zhifang Fang（Taiwan）
	罗清吉（台湾） 马晓燕 蒙小英 Hans Polman（荷兰） 邱坚珍 瞿 志	Zhe Huang Yuzhen Jian（Taiwan） Xiaoling Jin Chunfeng Li（Malaysia） Jianwei Li
	宋钰红 王明荣 王鹏伟 王秀娟（台湾） 吴静宜（台湾） 吴雪飞	Manliang Li Kaitai Lin（Taiwan） Chunqing Liu Tingfeng Liu Qingji Luo（Taiwan）
	吴怡彦（台湾） 夏海山 夏 岩 张莉欣（台湾） 张青萍 周武忠 周应钦	Xiaoyan Ma Xiaoying Meng Hans Polman（Netherlands） Jianzhen Qiu Zhi Qu
	朱 玲 朱卫荣 郑占峰 张新宇 赵晓平	Yuhong Song Mingrong Wang Pengwei Wang Xiujuan Wang（Taiwan）
		Jingyi Wu（Taiwan） Xuefei Wu Yiyan Wu（Taiwan） Haishan Xia Yan Xia
		Jianwei Ling Lixin Zhang（Taiwan） Qingping Zhang Wuzhong Zhou Yingqin Zhou
		Ling Zhu Weirong Zhu Zhanfeng Zheng Xinyu Zhang Xiaoping Zhao
编 辑	陈鹭 傅 凡 高 杰 孟 彤 佘高红 张红卫 张 安 赵彩君 马一鸣	Editors
	覃 慧（台湾） 郑晓笛	Lu Chen Fan Fu Jie Gao Tong Meng Gaohong She Hongwei Zhang An Zhang Caijun Zhao
		Yiming Ma Hui Qin（Taiwan） Xiaodi Zheng
外文编辑部	何友锋（台湾） Charles Sands（加拿大）（主任） Trudy Maria Tertilt（德国）	Foreign Language Editors
	谢顺佳（香港） 朱 玲	Youfeng He（Taiwan） Charles Sands（Canada, Director） TrudyMaria Tertilt（Germany）
		Shunjia Xie（Hongkong） Ling Zhu
版式设计	王 薇	Layout Design Wei Wang
责任编辑	刘恩桃	Editor-In-Charge Entao Liu
流程编辑	郭 真	Process Supervisor Zhen Guo
广告发行	宋焕芝 电 话：010-88364851	Advertising & Issuing Huanzhi Song Tel：010-88364851

地 址		Corresponding Address
北 京	北京市海淀区三里河路17号甘家口大厦1510	Beijing 1510A Room, Gan Jia Kou Tower, NO. 17 San Li He Street, Haidian District, Beijing P.R.C
	邮编：100037 电 话：86-10-88364851	Code No. 100037 Tel: 86-10-88364851 Fax: 86-10-88361443 Email: worldscape@chla.com.cn
	传 真：86-10-88361443 邮 箱：worldscape@chla.com.cn	HongKong
香 港	香港湾仔骆克道315-321号骆中心23楼C室	Flat C,23/F,Lucky Plaza,315-321 Lockhart Road, Wanchai, HONGKONG
	电话：00852-65557188 传 真：00852-31779906	Tel: 00852-65557188 Fax: 00852-31779906
台 湾	台北书局	Taiwan Taipei Bookstore
	台北市万华区长沙街二段11号4楼之6	6#,the 4th Floor , Changsha Street Section No.2, Wanhua District , Taipei Code No. 108
	邮 编：108 电 话：886+2-23121566，	Tel: 886+2-23121566 Fax: 886+2-23120820 Email: nkai103@yahoo.com.tw
	传 真：886+2-23120820 邮 箱：nkai103@yahoo.com.tw	Publishing Date July 2013
封面作品	法国布洛涅——比扬谷公园（图片来源：法国岱禾景观设计事务所）	Cover Story Parc de Billancourt, Boulogne–Billancourt, France Source : Agence Ter

图书在版编目（CIP）数据

世界园林．住宅景观：汉英对照/中国花卉园艺与园林绿化行业协会主编．
－－北京：中国林业出版社，2014.1
ISBN 978-7-5038-7340-9
Ⅰ．①世…Ⅱ．①中…Ⅲ．①住宅－景观设计－作品集－世界－现代Ⅳ．①TU986.61
中国版本图书馆CIP数据核字（2013）第321088号

中国林业出版社
责任编辑：李 顺 纪 亮
出版咨询：（010）83223051

出 版：中国林业出版社（100009 北京西城区德内大街刘海胡同7号）
印 刷：北京卡乐富印刷有限公司
发 行：中国林业出版社
电 话：（010）83224477
版 次：2014年1月第1版
印 次：2014年1月第1次
开 本：889mm×1194mm 1 / 16
印 张：12.5
字 数：200千字
定 价：定价：80.00RMB（30USD，150HKD）

简讯 NEWS

东方园林股票代码：002310

1992-2012

东方园林
20年
2000人
20座

城市景观艺术品
OrientLandscape
Urban landscape art

北京奥林匹克公园中心区景观
北京通州运河文化广场
首都机场T3航站楼景观
北京中央电视台新址景观
苏州金鸡湖国宾馆、凯宾斯基酒店景观
苏州金鸡湖高尔夫球场
上海佘山高尔夫球场
上海世博公园
海南神州半岛绿地公园
山西大同新城中央公园文瀛湖
湖南株洲新城中央公园神农城
辽宁鞍山新城景观万水河
辽宁本溪新城中央公园
河北衡水衡水湖及滏阳河景观
山东滨州生态景观系统及新城中心景观
浙江海宁生态景观系统
山东淄博淄河景观系统
河北张北风电基地及两河景观带
山东济宁微山湖及任城区中央景观
山东烟台夹河景观系统及特色公园

中国园林第一股
全球景观行业市值最大的公司
中国A股市场建筑板块、房地产板块前十强
城市景观生态系统运营商

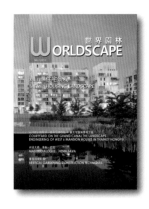

世界园林 第三辑
主　题 住宅景观

WORLDSCAPE
No.3 2013
THEME: HOUSING LANDSCAPE

WORLDSCAPE 目录

总编心语	014	
资讯	016	
	020	"园冶讲坛"成功举办
作品实录	026	运河岸上的院子泰禾红御西区6栋大宅园林景观工程 / 重庆天开园林
	034	荷兰卡佩勒市波德沙拉住宅区规划 / 安德斯建筑与都市设计：伊莲娜·舍甫琴科 & 肯托普森与戴维·莫瑞斯和安德鲁·肯斯金共同设计
	042	国城建设《高雄小城－一亩田》——一个生态小区的规划 / 洪嘉聪
	052	Quattro住区景观 / 宝克·科布东赛迪
	062	理性与感性的交融秦皇岛远洋海悦公馆展示中心景观设计 / 杨珂
	070	大一山庄住宅景观设计 / 广州高雅房地产开发有限公司
	081	法国布洛涅－比扬古公园　亨利·巴瓦　米歇尔·欧斯莱　奥利维耶·菲利浦
	088	滨海一号景观绿化工程 / 范美军　王彬彬　卢云慧　祁永
	094	无庶小区景观规划设计 / 江苏大千设计院有限公司
	104	贫民区的绿色生机温卡特中心花园的设计与建设 / 伊娃·卡纳普尔
对话大师	116	亨利·巴瓦
专题文章	136	论坡地住宅小区之景观意象及塑造 / 王小璘　何友锋
	144	株洲神农城核心区二期景观工程设计 / 李建伟
竞赛作品	156	渗透现象——呼和浩特市南湖公园湿地科普景区（方案一）/ 刘昶
	160	滋育东北农业大学校园广场景观规划设计 / 万亿
	164	哈尔滨市九站公园滨水区环境设计 / 王婧
新材料	170	垂直绿化工法 / 谢宗钦
征稿启事	185	
广告索引	封二	"园冶杯"住宅景观奖竞赛
	封三	中国风景园林网
	002	"园冶杯"风景园林（毕业作品、论文）国际竞赛
	005	北京东方园林股份有限公司
	008	棕榈园林股份有限公司
	009	深圳市柏涛环境艺术设计有限公司
	010	阿拓拉斯规划设计有限公司
	013	北京夏岩园林文化艺术集团有限公司
	023	盛世绿源科技有限公司
	024	北京市园林古建设计研究院有限公司
	112	上海亦境建筑景观有限公司
	113	济南市园林规划设计研究院
	114	无锡绿洲景观规划设计院
	152	重庆华宇园林股份有限公司设计分公司
	153	安道国际
	155	源树景观
	181	天开园林景观工程有限公司
	182	杭州市园林绿化工程有限公司
	183	盛世绿源科技有限公司
	184	广州山水比德景观设计有限公司
	186	浙江青草地园林市政建设发展有限公司
	188	北京海韵天成景观规划设计有限公司

WORLDSCAPE

CONTENTS

EDITORIAL 014
NEWS 016
    020   2013 "YUAN YE JIANG TAN" SUCCESSFULLY HELD

THE MASTER GARDEN
- **026** COURTYARD ON THE GRAND CANAL THE LANDSCAPE ENGINEERING OF WEST 6 MANSION HOUSES IN THAIHOT HONGYU / Tiankai Landscape
- **034** POLDER SALAD — RESIDENTIAL MASTERPLAN CAPELLE AAN DEN IJSSEL, THE NETHERLANDS
Anders architecture and urbanism: Elena Chevtchenko & Ken Thompson in collaboration with Dave Morison and Andrew Kitching
- **042** A DIFFERENT APPROACH TO COMMUNITY PLANNING / Chia-Chong Hong
- **052** A BEGGINING OF THE GARDEN—TILTING THE EARTH AND MARKING THE PLACE / Pok kobkongsanti
- **062** FLOW OF THE ORDER LANDSCAPE DESIGN FOR THE DISPLAY CENTER OF YUANYANG HAIYUE RESIDENTIAL AREA, QINHUANGDAO / Ke Yang
- **070** DAYI VILLA COMMUNITY / Guangzhou Gaoya Estate Development Co. Ltd.
- **081** PARC DE BILLANCOURT, BOULOGNE-BILLANCOURT, FRANCE
Henri Bava   Michel Hoessler   Olivier Philippe
- **088** GREENING AND ENGINEERING PROJECT IN BINHAI NO.1
Meijun Fan   Binbin Wang   Yunhui Lu   Yong Qi
- **094** LANDSCAPE DESIGN OF WUSHU RESIDENTIAL COMMUNITY / Jiangsu Daqian Design Institute Co.,Ltd.
- **104** A LITTLE BIT OF COUNTRY ON SKID ROW Designing & Building The Weingart Center Garden
Eva Knoppel

MASTER DIALOGUE
- **114** Henri Bava

ARTICLES
- **136** THE FORMING OF LANDSCAPE IMAGE FOR HILLSIDE RESIDENTIAL COMMUNITIES
Xiaolin Wang   Youfeng He
- **144** LANDSCAPE DESIGN FOR CORE AREA OF SHENNONG CITY IN ZHUZHOU, STAGE 2
Jianwei Li

WORKS FROMS
- **156** THE WATERFRONT LANDSCAPE DESIGN OF THE HOHHOT SOUTH LAKE PARK (PLAN I)
Chang Liu
- **161** MOISTEN AND NOURISH NORTHEAST AGRICULTURAL UNIVERSITY CAMPUS SQUARE LANDSCAPE PLANNING AND DESIGN / Yi Wan
- **164** THE WATERFRONT LANDSCAPE DESIGN OF THE NO.9 STATIONS PARK / Jing Wang

NEW MATERIALS
- **170** VERTICAL GARDENING CONSTRUCTION TECHNIQUES / Tzung-Chin Hsieh

185 Notes to Worldscape Contributors

ADVERTISING INDEX
- **Inside Front Cover** "Yuan Ye Award" Residential Landscape Competition
- **Inside Back Cover** www.chla.com.cn
- **002** The 2013 "Yuan Ye Award" International Landscape Architecture Graduation Student Design / Thesis Competition
- **005** Beijing Oriental Garden shares Co,Ltd
- **008** Palm landscape Architecture Co.,Ltd.
- **009** Botao Landscape
- **013** Xiayan Gardening Group of culture and art
- **023** ShengShiLvYuan Technology Co.,Ltd
- **024** Beijing Institute Of Landscape And Traditional Architectural Design And Research
- **112** Shanghai Edging A&La Co.,Ltd
- **113** Jinan Landscape Planning and Designing Research Institute
- **114** Wuxi Lv zhou Landscape Architecture & Plan Design
- **152** Chongqing Huayu Landscape &Architecture Co.,Ltd.
- **153** Aandi International
- **155** Yuanshu institution of Landscape Planning and Design
- **181** Tiankai Landscape Engineer Co.,Ltd.
- **182** Hangzhou Landscape Garden Engineering Co,ltd.
- **183** ShengShiLvYuan Technology Co.,Ltd
- **184** Sun & Partners Incorporation
- **186** Zhejiang Qingcaodi Garden Municipal Construction Co.,Ltd.
- **188** Macromind Architectural Landscape Planning & Design

Planning
Architecture
Landscape
Materialization

规划设计／建筑设计／景观设计／深化实施

PALM Design Co.,Ltd
风景园林设计甲级资质／建筑设计甲级资质／施工一级资质

长春柏翠园市民文化广场景观设计

广州
地址：广州市黄埔大道西638号
广东农信大厦18楼
邮编：510627
电话：020-37882900
传真：020-37882988

上海
地址：上海杨浦区淞沪路
388号创智天地7号楼三楼
邮编：200433
电话：021-31160001
传真：021-31160000

杭州
地址：杭州市江干区富春路
789号一楼
邮编：310008
电话：0571-89736005
传真：0571-89736006

北京
地址：北京朝阳区朝阳公园路19号
佳隆国际大厦B座21层
邮编：100125
电话：010-65390600
传真：010-65390700

成都
地址：锦江区东大街芷泉段
东方广场C座18楼
邮编：610021
电话：028-81453316
传真：028-84701136

景观规划设计·环境规划设计与咨询·产业研究·土地开发·地产策划·经济规划·可行性研究·城乡规划设计·旅游规划设计·建筑设计咨询·市政工程设计及相关技术管理与服务

棕榈园林股份有限公司
PALM LANDSCAPE ARCHITECTURE CO.,LTD.

黄山国际中心

黄山国际中心，该项目荣获2013年第八届金盘奖全国最佳商业楼盘。项目位于安徽省黄山市屯溪区东南角黎阳镇内，黎阳古镇始建于东汉建安年间，距今1800年整，1000多米长的黎阳古街颇具韵味。景观设计以黎阳老街改造为核心，以符合现代生活方式的休闲商业广场为配套，突出徽州本地历史文化传承，并赋予其新的形象语言。

创新是设计之魂，质量是立足之本，服务是发展之道
INNOVATION IS THE SOUL OF DESIGN, QUALITY IS THE BESE, SERVICE IS THE WAY OF DEVELOPMENT

深圳市柏涛环境艺术设计有限公司　BOTAO LANDSCAPE ART & DESIGN CO.,LTD
地址：深圳市南山区华侨城生态广场C栋101　电话：0755 26919715　26913889　网址：www.botaoead.com　新浪微博：柏涛环境

ATLAS アトラス | 10th 阿拓拉斯十周年

中国　电话：00
北京市海淀区知

常务理事单位

无锡绿洲景观规划设计院有限公司

EDSA Orient

北京大元盛泰景观规划设计研究有限公司

北京夏岩园林文化艺术集团有限公司

棕榈园林股份有限公司

岭南园林股份有限公司

北京源树景观规划设计事务所

北京欧亚联合城市规划设计院

重庆金点园林股份有限公司

重庆天开园林景观工程有限公司

杭州天香园林有限公司

北京市园林古建设计研究院有限公司

盛世绿源科技有限公司

理事单位

北京东方园林股份有限公司

重庆华宇园林股份有限公司设计分公司

枫彩农业科技有限公司

江苏山水建设集团有限公司

苏州新城园林发展有限公司

广东四季景山园林建设有限公司

北京乾景园林股份有限公司

广州市林华园林建设工程有限公司

夏岩文化艺术造园集团
XIAYAN PLANNING & LANDSCAPE

岩二勤生态园
生态餐厅设计建造

天津龙达温泉生态城
温泉水城设计建造

淹城诸子百家园
主题公园设计建造

Add：北京市朝阳区高井文化产业园8号东亿国际传媒产业园 C7号楼6层
www.xiayan.cn

垂询热线 400-675-5886

北京/沈阳/天津/常州/南京/新疆

总编心语 / EDITORIAL

王小璘
Xiaolin Wang

住宅景观设计的人本主义与住宅环境的生态可持续是当前住宅景观设计中的热门议题。住宅景观设计的人本主义强调在满足人们物质需求的同时，亦关注垫基于个别住宅环境的社会福祉；其意涵主要包含两个层面，一是强调对不同人群使用需求的重视，发挥景观及其生态和社会的功能；二是通过景观设施、小区互动环境和无障碍空间等的人性化设计，提升小区居民的归属感，促进小区的社会活力。另一方面，绿色能源和生态建筑技术以及诸多节能减排手段，已经成为人居环境建设的重要课题，住宅景观的可持续性不仅是单纯的资源节约，更在于恢复居住区的生态调控功能，乃至构成城市的有机组成部分，进而达到城市生活、生产、生态三生功能整合的最终目标。

近年来，随着城镇化和城市基础建设不断扩大，居住区已成为城市环境重要的一环，人们的择居理念亦由"居者有其屋"发展到"居者择其屋"和对更高质量生活空间与景观环境的追求。面对于此，风景园林师亟思如何在宏观与微观各个层面去拓展可持续且宜居的住区景观设计。

本期共介绍荣获景观设计首奖的七个作品，包括来自中国大陆「2012年园冶杯住宅景观奖」，台湾「2013年高雄市透天建筑景观类建筑园冶杯奖」等奖项及欧洲「青年建筑师竞赛」以及美国「2012年景观设计师协会住宅设计类竞赛」等以及四个优秀作品。专题方面则以景观意象之观点论述坡地住宅小区之型塑，为住宅景观提供不同面向的设计思维。

本期焦点人物为法国景观设计师亨利巴瓦教授。作为法国岱禾景观建筑设计事务所的首席设计师，巴瓦教授的作品已遍及许多欧洲及其它国家；其中，2010上海世界博览会法国馆即为其代表作之一。通过对话大师一席谈话，读者可以一窥这位严谨而有深度，理性和感性兼俱的景观大师如何看待景观设计场域所隐藏的内涵，从而透过细腻的设计思维和手法，展现每个作品不同的特色。

「垂直绿化工法」系本刊首次推出介绍的「新工法」。在都市开放空间和公园绿地逐渐为建筑用地所取代的趋使之下，垂直绿化成为弥补都市绿地和提高绿量的国际趋势。本刊盼能抛砖引玉唤起城市规划和建筑设计对此一课题和技术的重视。

王小璘

In the field of residential landscape design, the two issues currently garnering the most attention are humanism and ecological sustainability. Humanism in residential landscape design emphasizes the satisfaction of human needs and the promotion of social welfare from the basis of the individual's residential environment. This is pursued on two main levels, one is to pay attention to the needs of different social groups in relation to landscape and its ecological and social functions. The other is to enhance human-based residential landscape design with landscape facilities, environments for community interaction, and more inviting spaces, for the purpose of building a greater sense of community and promoting a more expansive field of social dynamics. On the other hand, green energy resource technology and eco-construction techniques, along with other means of reducing energy consumption have become a focus of residential community construction. Although resource conservation is not an issue unique to residential landscape sustainability, ecological restoration is more imperative in residential areas. It is a matter of coordinating the ecological benefits of urban green space with the larger organic processes of a city. Thus the ultimate goal of uniting the ecological, economic and social aspects of the environment can be achieved.

In recent years, urbanization and infrastructure have been continuously explored. Along with this, a greater focus has been placed on the role of residential areas in the optimization and renovation of the urban environment. At the same time, the individual's residential priorities have transitioned from simply 'home ownership', to ownership of the ideal type of housing, to the pursuit of a higher environmental quality of life. Facing to this trend, landscape designers should consider how to explore sustainable and livable residential landscape design at both the macro and micro scales.

In this issue we introduce seven outstanding projects, including: the first prize of 'The 2012 Yun-ye Awards of Residential Landscape Competition' in China; the 2013 Kaoshung Architectural Yun-Ye Cup Landscape Category Competition' in Taiwan; 'The European Young Architect Competition'; and the 2012 Residential Design Competition of the American Society of Landscape Architects', and four outstanding projects. In the 'Articles' section, the formulation of residential environmental phenomenon based on the concept of landscape image is introduced. This landscape design philosophy is then applied towards the design of hillside residential areas.

As chief landscape architect of 'Agenceter' in France, Professor Henri Bava's works spread across Europe and the world. The French Pavilion from the 2010 International EXPO stands out as one of his most representative projects. In the 'Master's Dialogue' section we will gain insight into how this rigorous, deep, logical and romantic master landscape architect approaches the landscape context to create his unique and varied designs.

The Vertical Greening Technique' is the first 'new method' addressed in this publication. Due to the pressure of urban development, public open space, parks and green land have been threatened by new construction. As part of the goal of increasing green mass and coverage in the urban environment, vertical greenery has become an international trend. We hope our exploration of the techniques involved with vertical greening will prove useful for urban planners and architects.

资讯 NEWS

澳大利亚"2020环保城市计划"

随着越来越多的人移居城市，城市的自然植被渐渐枯萎，人们居住空间中绿色的植物越来越少，这样下去，迟早有一天，城市中将没有树木来净化空气、提高环境质量。澳大利亚政府正打算改变这一现状。澳大利亚政府、学术和私营部门协力推出的"2020愿景"计划，旨在为澳大利亚城市中心创造百分之二十的绿化面积。由于绿色空间不被纳入城市的新发展和大型的建筑项目，城市热岛效应、空气质量差、良好的城市社区环境的缺乏等都是产生的不良结果。

来源：中新网

丹麦BIG工作室设计的LEGO之家（LEGO House）

丹麦BIG工作室设计的LEGO之家（LEGO House）将作为乐高之家这款世界著名积木玩具品牌的游客接待中心。乐高之家位于乐高玩具的故乡，丹麦比隆（Billund）。建筑本身就像一座放大了的乐高玩具，由大小不同的体块堆积而成。建筑内设置了商

店、展厅以及各种体验空间。建筑的屋顶等部分作为开放空间，成为市民休闲娱乐的空间。人们仿佛变成了LEGO世界的小人，在LEGO世界畅游。

LEGO之家将于2014年动工建设，预计2016年落成。

来源：灵感日报

White Arkitekter赢得美国Resilient Rockaways设计大赛冠军

2013年的国际设计竞赛中，斯德哥尔摩的景观公司White Arkitekter赢得Resilient Rockaways设计大赛冠军。大赛的初衷是因为2012年10月"桑迪"飓风造成纽约地区沿海广泛遭遇损害，因此需要增强弹性基础设施，并提高现有的沿海社区设施，这是大纽约州地区重建策略的需要。该设计源于引入一个设计理念，采用小的变化和策略，利

用适应性强、可行的，经济实惠、智能的方法，达到最终的维护目标。从基础来说，该设计有助于保护海岸沿线，利用天然沙滩制作风暴潮水缓冲区。

来源：筑龙网

迪士尼将在地球上建立潘多拉阿凡达世界

如果你曾深深迷恋詹姆斯卡梅隆镜头下的阿凡达世界，这个梦想很快就能成真。迪士尼公司宣布，将计划在地球上建立阿凡达星球。该世界将会充满电影中的多彩的植物、悬浮在空中的岛屿、大量的发光的动植物等等。新园区将于2014年在佛罗里达州动工，2016年开放后，在园中开拍阿凡达续集。

来源：筑龙网

意大利：Neighborhood社区公园

这是圣多纳迪皮夫城市中一个小型的社区公园，这个社区没有多少公共用地，仅有的空地也存在可达性不高的问题。这个社区公园的出现改变了这一不良现状。社区公园的南部面向主要公路，内部的铺装是白色的水洗石，这白色的铺装如同自然地貌般起伏着，形成小山、盆地，还有聚会的座椅、喷泉、道路、儿童乐园、野餐点、树木、灯具，

以及被树木穿过的白色石凳帮助定义着公园的空间,并在不同的季节为人们带来不同的小气候。

整个公园连续而统一,白色的路面串起所有的功能,夜晚来临,修长的灯杆泛光灯源照射着下方柔和的路径。如今这里变成一个非常热闹的聚会场所,全天候的迎接各种各样的人们,有读书看报的,有骑山地车的,有闲聊的,还有玩音乐的。

来源:谷德设计网

2017年阿斯塔纳世博会入围方案

2017年世博会的主题为"未来的能量"("The Energy of the Future")。Saraiva + Associados建筑事务所为阿斯塔纳地区打造出一个具有恒久意义的地标建筑,通过整合可再生资源、提高能源效率、使用清洁技术,并依托可持续发展的原则使城市更加宜居。

本次博览会将在阿斯塔纳左岸构建一个逻辑结构,这样的城市规划场景符合最有利于城市发展的可持续发展的"保利中心"原则。设计充分考虑了当地夏季温暖冬季寒冷的特殊气候条件。两个复杂的轴线以90度的弧线相交叉,这些轴线都是南北向和东西向的,靠近交叉点并且位于世博的象征处。它在标志性建筑物之间和拜特雷克观景塔之间建立了一个强大的关系。最重要的是从这里打开了仪式广场。这一复杂的建筑采用的是世界级的可持续技术,这样的设计是想尽量让更多的人在这个未来城市中参与到这次盛会。

来源:中国建筑报道网

瑞典于默奥大学公园景观

默奥大学是一所成立于20世纪60年代后期的年轻大学。在这里,来自全国各地的35000名学生进行着各个知识领域的研究。大学坐落于北极圈南部约300公里左右的海岸。

一所大学应是兼容多元化思想沟通的场所。它并非像礼堂或实验室那样,而是无等级的开放空间,研究人员及师生可以在此互动交流。从而大大提升校园整体水平和吸引力。

默奥大学新校区公园由23000平方米的阳光甲板、码头、开阔草坪、步行道和人工湖及其周围用来举办活动的露台组成。湖心岛处有栈道通往湖的南岸。在这里,游客可以同时欣赏到阳光明媚的丘陵景观以及白桦树干斑驳交织的光影幽谷。

来源:筑龙网

欧洲最大图书馆:改建后的英国伯明翰图书馆

伯明翰图书馆是一个透明的玻璃建筑。其精巧细腻的皮肤灵质感来自于这个以前的工业城市的工匠传统。电梯和自动扶梯动态地置于图书馆中心,形成空间内八个圆形建筑之间的连接。这些圆形大厅发挥重要作用,不仅作为图书馆的通路,同时也提供了自然采光和通风。图书馆体量之间的错叠悬挑形成了屋面露台,屋面露台被设计成为美丽的屋顶花园,是城市中的大阳台。

来源:土木在线

英国国王十字火车站野餐公园

英国伦敦国王十字火车站一个隐秘的角落被改造成为极富魅力的野餐公园。公园内一个标志性的铁路墙是伦敦建筑节的参展作品之一,由当地Squireand Partners建筑事务所联合景观设计师Jeremy Rye及艺术家Anna Garforth.共同创作设计。

周末,来自各地的500名游客来这里野餐,这里有铺满草皮的躺椅,上面点缀着盛开的本土鲜花和花篮,铁路墙上设计了以苔藓为原材料的图案模纹,它为公园提供了一个引人注目的背景。公园的大部分设施都是由回收的画板制作而成,儿童们可以拿着粉笔任意涂画。

内部空间则设计成为维多利亚仓库餐厅,餐厅里布置着更多的草皮,定制的游戏桌和座椅也是由旧画板制作而成。草皮的上方,人们可以享受在室内荡秋千的乐趣。

来源:筑龙网

国际重要湿地和国家湿地公园"扩编"

10月24日,在山东省东营市召开的第三届中国湿地文化节暨东营国际湿地保护交流会议议程中,我国新增加5块国际重要湿地,20个湿地公园通过验收正式成为国家湿地公园。至此,我国国际重要湿地总数达46处,国家湿地公园总数达32个。

新增的5块国际重要湿地是:山东黄河三角洲国家级自然保护区、黑龙江东方红湿地国家级自然保护区、吉林莫莫格国家级自然保护区、湖北神农架大九湖湿地和武汉沉湖湿地自然保护区。

来源:中国林业网

英国伦敦都市绿洲

这是由 KHBT / OSA_OFFICE FORSUBVERSIVE ARCHITECTURE 设计的英国伦敦都市绿洲。Hoegaarden 都市绿洲城市是一个临时的户外设施,它挑战了城市和自然环境事件的关系,同时城市居民们对于它的全新的功效也有着新的反馈。与 Hoegaarden 都市绿洲一起,OSA 在一个城市中心创造了一天绿色的空间,这个设施从生理上和精神上,都为都市人们创造了一个绿洲,当人们进入、互动、然后可以从城市紧张的生活中放松身体和精神。这座绿洲被安装在一个酒吧的顶部,设计成为草坪地毯的形式,采用体育场的线条,制作了 60 个座位区和小型吧台。

来源:筑龙网

第十一届景观教育大会(地理设计)在京召开

10月28日至29日,"北京大学第十一届景观设计学教育大会暨2013年地理设计国际会议"在北京大学英杰交流中心举行。本次大会由北京大学建筑与景观设计学院和 ESRI 中国信息技术有限公司共同主办,景观中国网站和《景观设计学》杂志承办。为期两天的会议分两个部分:围绕"地理设计:人地关系优化设计的理论与实践"的主旨报告环节和分别以"协同探索——地理设计的应用与创新"和"协同工作——地理设计实践前沿"为主题的两个分论坛。

来源:景观中国网

俄罗斯:加大力度保护文化遗产

日前,俄罗斯总统普京签署《俄罗斯联邦行政违法法典》修正案,大幅提高毁坏历史文化遗产的罚金数额。此外,俄罗斯还设立了专门的历史文物保护区,在这些区域内不得建设新项目,否则就是违反法律,这些历史文物保护区由俄文化部下属的文物保护局统一管理,一旦列入《世界遗产名录》,就由俄罗斯遗产委员会负责保护。

在俄罗斯有这样一句口号:保护文化遗产就是保护俄罗斯母亲!俄罗斯民族创造了光辉灿烂的文化,每个俄罗斯人都为此感到自豪。

来源:光明日报

澳大利亚:Box Hill 运动公园

澳大利亚墨尔本 Box Hill 运动公园历史悠久,墨尔本市委托 ASPECT Studios 将其设计成为多样休闲娱乐活动的花园,创造了一个新的公共空间。

项目的主要任务是改造1个网球俱乐部的结构。部分会所被保留,休息平台也再利用。新的厕所位于翻新的墙后。

该项目作为社区公共绿色空间和多种体育和娱乐活动的场地,利用充满活力的图形,

重新定义了游乐区和日益增长的社区需求,成为一个标志性而又活泼的地点。在未来,1公里长的步行和跑道将会被增建。这条路径将会环绕花园,起点和终点连接,成为新的多用途场地。

来源:筑龙网

布满"鱼鳞"的建筑:墨西哥城 Soumaya 博物馆

图为墨西哥城的 Soumaya 博物馆,这是一家私人拥有的博物馆,运营者是 Carlos Slim 基金会。其自由形状的外观是由 FREE Fernando Romero EnterpriseE 设计的,展览空间为6万平方英尺,坐落在原先的一个工业区内。这座博物馆是一个有机的多维建筑,不透明的外墙用大量六边形的铝制模块搭建,提高了建筑的经久耐用。

Soumaya 博物馆的设想是一个有机的和不对称形状的旋转雕塑块。建筑的独特外观是由六角形铝模块构成的,这是为了方便的保存和耐久性。soumaya 博物馆的设计意图是在形态学和类型学之间建立一个桥梁,它定义了新的范例模式。

来源:土木在线

园博会闭幕式报道

  2013年11月18日，历时半年的第九届中国（北京）国际园林博览会圆满闭幕。当天下午，第九届园博会闭幕式暨颁奖大会在北京隆重举行，参会人员主要有住建部、中国风景园林学会、中国公园协会领导及北京市委、市政府、市人大、市政协领导，各省、自治区住建厅和园林绿化局代表、国内各参展城市代表及建设单位代表，，组委会成员单位代表及新闻媒体代表总共约350人，共同见证了北京园博会的完美谢幕。

  会议开始，参会人员共同观看第九届园博会专题片，回顾了北京园博会的历程与辉煌。而后，建设部领导宣读表彰文件、颁奖，向志愿者献花表示感谢。

  颁奖仪式完毕后，住建部领导宣布第十届园博会举办城市为武汉，北京、武汉两地代表对交接会旗。

  中国风景园林网作为北京园博会大师园和设计师广场园区的策划组织方，其工作受到组委会的肯定和鼓励，被授予"第九届中国（北京）国际园林博览会先进集体"称号，《世界园林》杂志编辑宋焕芝被授予"第九届中国（北京）国际园林博览会先进工作者"称号。

  统计显示，自5月18日开幕以来，北京园博会共接待游客610万余人次，日均接待3.3万余人次，单日最高游客接待量10.6万人次，均创历届园博会之最。

2013"园冶讲坛"成功举办
"Yuan Ye Jiang Tan" Successfully held

2013年"园冶杯"风景园林(毕业作品、论文)国际竞赛(简称"园冶杯"竞赛)在组委会的统一组织下,竞赛评审顺利进行。由中国风景园林网和世界园林杂志社共同主办的"园冶讲坛"系列活动也随着"园冶杯"竞赛评审的开展拉开了序幕。获奖作品的创新思维和作品质量的高水准,受到评委专家的一致好评,本次"园冶讲坛"也在风景园林学界和广大高校的师生间引起了极大的反响。

2013年度"园冶讲坛"系列讲座共分三场进行,分别在北京农学院、北方工业大学和南京林业大学举行。台湾造园景观学会名誉理事长、《世界园林》杂志主编王小璘教授、南京林业大学风景园林学院张青萍副院长、华南农业大学风景园林与城市规划系主任李敏教授等30个国内外高校的教授和设计院的40余名知名设计师做出了精彩的报告。有上千名学生参与了这次论坛,在广大院校的师生中间引起了极大的反响。

南京林业大学风景园林学院副院长赵兵教授主持论坛

台湾造园景观学会名誉理事长 王小璘教授

华南农业大学系主任 李敏教授

东南大学建筑学院景观系主任 成玉宁教授

四川农业大学 院长 陈其兵

中国农业大学观赏园艺与园林系副主任 孟祥彬教授

北京海韵天成景观规划设计有限公司首席设计师 顾志凌

广州山水比德景观设计公司总经理兼设计总监 孙虎

南京赛诺格顿景观设计公司设计总监 张增记

第二论坛现场

中南林业科技大学风景园林学院副院长胡希军教授演讲

园冶讲坛旨在推动各院校间的相互交流，促进中国风景园林教育的和谐发展，同时为不同地区的学生提供学习交流的平台。

演讲进行过程中会场座无虚席，在场的师生及设计师们仔细聆听老师的讲座，整个过程精彩而有序。

2013"园冶讲坛"在举办期间受到了各高校师生与业内人士的支持与肯定。"园冶讲坛"系列活动旨在促进学术交流，为各高校广大师生提供一个交流思想、互相学习的平台。今后，"园冶讲坛"会在更多的院校举办，让"园冶杯"竞赛的热潮在高校间继续传递，推动风景园林事业的发展。

南京林业大学风景园林学院副院长张青萍　　EDSA总裁、首席设计师李建伟

第三论坛现场

"园冶杯"
青年
风景园林师
奖

报名 / 作品提交时间： 2014 年 5 月

提交材料： 申报书、个人作品资料和获奖项目证明材料（按 A3 标准装订成册）一套（并附作品光盘）；鉴于评审工作的需要，另请提供介绍申报者的演示光盘（限 10 分钟，演示光盘免除背景音乐）

申报条件： 严格遵照《园冶杯青年风景园林师奖申报及评审条例》中规定的申报条件及要求

主办单位： 园冶杯风景园林国际竞赛组委会

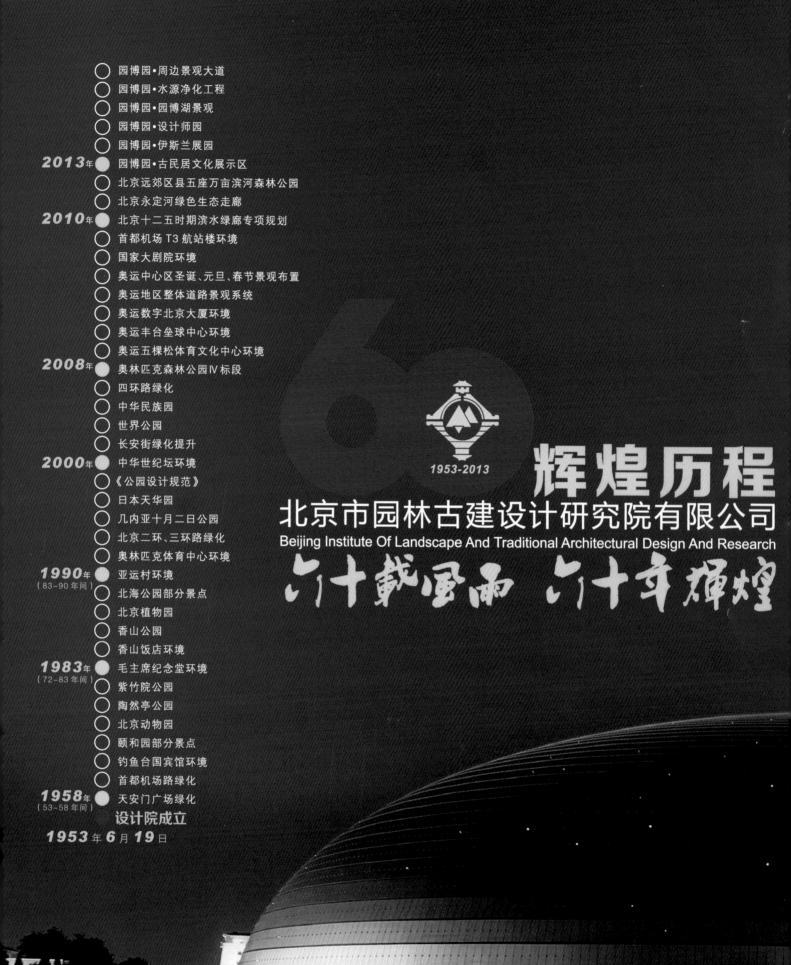

辉煌历程

北京市园林古建设计研究院有限公司
Beijing Institute Of Landscape And Traditional Architectural Design And Research

六十载风雨　六十年辉煌

1953-2013

- 2013年
  - 园博园·周边景观大道
  - 园博园·水源净化工程
  - 园博园·园博湖景观
  - 园博园·设计师园
  - 园博园·伊斯兰展园
  - 园博园·古民居文化展示区
  - 北京远郊区县五座万亩滨河森林公园
  - 北京永定河绿色生态走廊
- 2010年
  - 北京十二五时期滨水绿廊专项规划
  - 首都机场T3航站楼环境
  - 国家大剧院环境
  - 奥运中心区圣诞、元旦、春节景观布置
  - 奥运地区整体道路景观系统
  - 奥运数字北京大厦环境
  - 奥运丰台垒球中心环境
  - 奥运五棵松体育文化中心环境
- 2008年
  - 奥林匹克森林公园Ⅳ标段
  - 四环路绿化
  - 中华民族园
  - 世界公园
  - 长安街绿化提升
- 2000年
  - 中华世纪坛环境
  - 《公园设计规范》
  - 日本天华园
  - 几内亚十月二日公园
  - 北京二环、三环路绿化
  - 奥林匹克体育中心环境
- 1990年（83-90年间）
  - 亚运村环境
  - 北海公园部分景点
  - 北京植物园
  - 香山公园
  - 香山饭店环境
- 1983年（72-83年间）
  - 毛主席纪念堂环境
  - 紫竹院公园
  - 陶然亭公园
  - 北京动物园
  - 颐和园部分景点
  - 钓鱼台国宾馆环境
  - 首都机场路绿化
- 1958年（53-58年间）
  - 天安门广场绿化
  - 设计院成立
- 1953年6月19日

A unified vision
To the visual
成就非凡 卓识远见

北京市园林古建设计研究院有限公司初创于1953年,是我国最早从事风景园林设计的单位之一,是第一批经建设部批准的"风景园林"甲级设计资质单位,同时拥有建筑工程乙级设计资质,具有规划咨询、园林设计、建筑设计等综合设计实力,可承揽相关领域的规划设计任务。

北京市园林古建设计研究院有限公司拥有一支集风景园林师、规划师、建筑师以及结构、给排水、电气、概预算等多专业工程师组成的160余人综合设计团队,其中拥有中高级技术职称人数近60%。

六十年来,**北京市园林古建设计研究院有限公司**凭借自身实力以及丰富的实践经验,始终处于行业中的领先地位,设计成果遍及国内外,如颐和园耕织图景区复建工程、北京奥林匹克森林公园Ⅳ标段施工图设计、国家大剧院景观工程、德国得月园、日本天华园等都得到了行业主管部门和社会人士的一致好评;多次在国家级、部级、北京市优秀设计和科技进步奖评选中获奖,累计达140余项。

北京市园林古建设计研究院有限公司是中国勘察设计协会常务理事单位、中国勘察设计协会园林与景观设计分会副会长兼秘书长单位、北京市勘察设计协会常务理事单位、中国风景园林学会理事单位、北京市园林学会理事单位,主持并参与了多项行业标准的制定,自行研发的园林规划设计软件在行业中广泛应用。

地址:北京市海淀区万寿寺路6号 电话:010-68423979 / 68423969 网址:www.ylsj.cn

作品实录 / PROJECTS

运河岸上的院子
泰禾红御西区 6 栋大宅园林景观工程

COURTYARD ON THE GRAND CANAL
THE LANDSCAPE ENGINEERING OF WEST 6 MANSION HOUSES IN THAIHOT HONGYU

重庆天开园林　　　　　　　　　　Tiankai Landscape

项目位置：通州北京通燕高速宋庄出口南行 800 米京杭大运河畔
项目面积：24,000m²
委托单位：北京泰禾房地产开发有限公司
施工单位：重庆天开园林股份有限公司
景观设计：张永和
完成时间：2011 年 12 月
获　　奖：2012 年园冶杯住宅景观奖工程金奖

Location: Tongzhou, Beijing
Area: 24,000m²
Client: Beijing Thaihot Real Estate, Co. LTD
Designer: Palm Landscape Architecture Co. LTD
Landscape Design: Yonghe Zhang
Completion: December, 2011
Awards: 2012 Yuanye Engineering Gold Cup Award for Residential Landscape

01

图 01 庭院式住宅
Fig 01 Courtyard house

图 02 庭院式住宅
Fig 02 Courtyard house

图 03 欧式建筑，别样风味
Fig 03 Residence in the European style

图 04 生态自然
Fig 04 Ecologically friendly

图 05 绿意盎然
Fig 05 Abundant with greenery

图 06 生机勃勃的古典园林
Fig 06 The lively traditional of Chinese classical gardens

工程介绍：

　　我公司承建的泰禾红御西区6栋大宅园林景观工程（图01-03）位于长安街东起点，坐落京杭大运河河畔，北京通燕高速与长安街一线相通，仅20分钟就可以顺畅直达CBD，坐拥京杭大运河的原生水岸，天然的生态景观（图04-05），使该工程成为真正的第一居所城市纯独栋院落别墅。每座大院（图06）的庭院面积都大于建筑面积，无论从景观营造还是生活的空间尺度，都无一不突出景观的自然与融合，更能体会到欧式建筑的别样风味。庭院中天开地阔，可植树栽花（图07），饲鸟养鱼，将自然融入生活，把自然拉进人心；而建筑融入庭院，主人的空间能够一直延伸到私家庭院之中，将生活的范围和足迹从室内空间释放出来，室内与庭院空间相交融（图08-10），使得居住环境变得开放流通，庭院的设计给室内多个空间带来了丰富的对景景观，使庭院成为了视线的焦点与落点，人与自然也更加贴近。该工程的庭院定位于现代欧式风格私家别墅，在吸收了传统欧式风格以水景为风

The introduction to the project

　　The 6 mansion houses in the west Thaihot Hongyu start from west Changan Avenue and stretch along the Grand Canal from Beijing to Hangzhou (Fig 01-03). Since the Tongyan Expressway is connected to Changan Avenue, it only takes 20 minutes to reach the central business district. The natural ecological scenery along the Grand Canal enables the project to be virtual detached housing (Fig 04-05). The size of the courtyard of each house is larger than that of the house (Fig 06). The combination with nature embodied in the landscaping and living spaces guides the residents to these European style residences. The yards are large enough to plant flowers (Fig 07), feed birds and fish, and get close to nature. It opens a dialogue with the surrounding space by introducing a new way of community living, mainly based

直树栽花，自然融入生活
Nature comes to life when trees and flowers planted

图 09 生活的范围和足迹从室内解放出来
Fig 09 Life isn't limited to the indoor area

图 10 主人的空间一直延伸到庭院之中
Fig 10 The living space stretches to the outdoors

08 室内空间与庭院相交融
08 The integration of the indoor space and the outdoors

图 11 精致的凉亭与周围景色遥相呼应
Fig 11 Exquisite pavilion

图 12 景观小品清新雅致
Fig 12 Outstanding views

图 13 水景唤醒了阵阵生机
Fig13 The water landscape

图 14 花坛点缀的盆式小喷泉
Fig 14 Fountain ornamented by tiny flower beds

图 15 花坛中鲜花绚丽盛开
Fig 15 Flowers in bloom

图 16 中国古典庭院整齐典雅
Fig 16 A Chinese traditional courtyard

图 17 品质在于细节
Fig 17 It is the details that count

图 18 古典雅致
Fig 18 Classical and graceful

图 20 静谧的内心和安静休闲的庭院相呼应
Fig 20 The tranquil courtyard is the reflection of the quiet mind

图 19 日常生活的艺术化
Fig19 Daily life can be artistic

情特点的同时，融入了中式园林施法自然的配景手法。庭园中常常布置装饰性建筑，如柱廊、圆亭、凉亭（图11）、观景楼、装饰景墙等。由于后院为下沉式庭院，建筑多依坡势而建，庭园侧面开辟出"梯田"式的台地，种植高耸的蒙古栎，元宝枫等大树，而平台、花坛、雕塑等小品多对称布置（图12）。庭园主景多是在中轴线的宽路上设置雕塑、水景（图13-14）或花坛（图15），水景采用盆式的小喷泉。造景方面则尤其讲究借景与园外的自然环境相融合，注重乔木灌木以及花卉的层次搭配和栽植方式，大气之中不乏细节之处的细腻经典。

项目理念

泰禾红御西区6栋大宅园林景观工程作为运河大宅的代表之作，自甲方到我公司都极为重视，经双方领导多次接触，最终确立结合当地地脉价值、人文底蕴，回归中国人千百年的院居生活情结，打造中国大院欧式园林景观（图16-18）。

项目效果

泰禾庭院的造园手法，基于我公司"虽由人作 宛自天开"的朴素思想，以G07为例，透露出处处师法自然，塑造从物到心的三重境界：一是"物境"，通过雕塑、水景、花坛、柱廊、圆亭、装饰景墙等元素的随意运用，在庭院内部还欧式景观之本；二是"情境"，庭院随四季交替而展现出现出丰富的季相，我们使景致融合心境，体会欧式风格之丽；三是"意境"，我们将庭院景致日常生活艺术化（图19-20），烛光晚宴、草坪聚会、露天音乐会，一切活动既是内心对庭院的呼应，又使庭院这个空间背景具有诗一样的意境。以"群体联络之美"著称的泰禾院落，在庭院景观结构上处处追求收放对比，层层递进，视线上的先抑后扬，让潜意识中会觉得空间无限放大。庭院景观构造不拘一格，结合水岸、地势、密林巧妙搭配的植物造成间隔，无形间把庭院自然分隔，动线曲折，营造空间"由外部开放到最私密空间再到个人开放"的丰富变幻，而整个植物搭配过程始终围绕庭院展开（图21），"步移景异"，游不完的院子，看不尽的美景。□

on the generosity of the exterior private and community spaces (Fig 08-10). The designer's choice of opposing landscaping techniques and views in the courtyards brings the inhabitants closer to nature. The project aims at the modern European style of detached houses, and it combines the water-landscape-featured traditional European style with the Chinese nature-featured object-based style. Decorative constructions are seen in each of the yards such as pavilions (Fig 11), gazebos, and decorative walls. As the backyard is in the sink-style, the buildings are largely constructed along slopes. A terraced platform is built on one side of the courtyard to plant high trees like Mongolian oak(*Quercus mongolica*) and Purpleblow maple(*Acer truncatum*)(Fig 12). The main features like the statues (Fig 13-14), the water landscape and the flower beds are set on an axis to the courtyard while the decorations are symmetrically set (Fig 15). The garden design focuses on the integration of nature and borrowed scenery, and the leveled planting of trees and flowers: All of these elements embody both the grandness and the detail of the landscape.

Project Concept

As the representative of the residence along the canal, the project has been paid great attention by both the client and us. We finally reach an agreement on constructing the residence into a western landscape with Chinese characteristics, where the residents can enjoy a courtyard life that has lasted for hundreds of years.

The greatest achievement of the project: the garden-making skills, are based on our philosophy of "The masterpiece of nature, through artificial gardens". There are three conceptions when

图 21 整个植物搭配始终围绕庭院展开
Fig 21 Courtyard-oriented plantings

following the principles of nature. The first is the voluntary application of the conception of concrete elements such as statues, water landscape, follower beds, pavilions and decorative walls; The second is the conception of emotions that are brought by the changing of seasons; The Third is the conception of the artistry of daily activities (Fig 19-20). Renowned for "the connected beauty of clusters", the Thaihot courtyard tries to achieve a progressively larger effect of views. The landscape structures are of different genres but are tied together through the integration of privacy and openness of communication (Fig 21), creating a multi-element space featuring a variety of landscapes.■

作者简介：

重庆天开园林股份有限公司 / 园林景观设计与施工 / 中国重庆

Biography:

Tiankai Landscape / Garden landscaping and engineering / Chongqing, China

查尔斯·沙（校订）
English reviewed by Charles Sands

荷兰卡佩勒市波德沙拉住宅区规划
POLDER SALAD — RESIDENTIAL MASTERPLAN CAPELLE AAN DEN IJSSEL, THE NETHERLANDS

安德斯建筑与都市设计：伊莲娜·舍甫琴科 & 肯托普森与戴维·莫瑞斯和安德鲁·肯斯金共同设计

Anders architecture and urbanism: Elena Chevtchenko & Ken Thompson in collaboration with Dave Morison and Andrew Kitching

图01 总平面图
Fig01 Master Plan

项目位置：卡佩勒·安·丹·艾瑟尔，阿姆斯特丹，荷兰
项目面积：50,000m²
委托单位：卡佩勒·安·丹·艾瑟尔政府
设计单位：安德斯建筑景观规划设计
景观设计：伊莲娜·舍甫琴科，肯托普森
建筑设计：伊莲娜·舍甫琴科，肯托普森，戴维·莫瑞斯，安德鲁·肯斯金
完成时间：2015年（一期）
获　　奖：欧洲青年建筑师竞赛一等奖

Location: Capelle aan den IJssel, Rotterdam, The Netherlands
Area: 50,000m²
Client: Municipality of Capelle aan den IJssel, Havensteder
Designer: Anders architecture masterplanning landscape
Landscape Design: Elena Chevtchenko, Ken Thompson
Architecture Design: Ken Thompson, Elena Chevtchenko, Dave Morison, Andrew Kitching
Completion: 2015 (phase 1)
Awards: The first prize in the European competition for architects

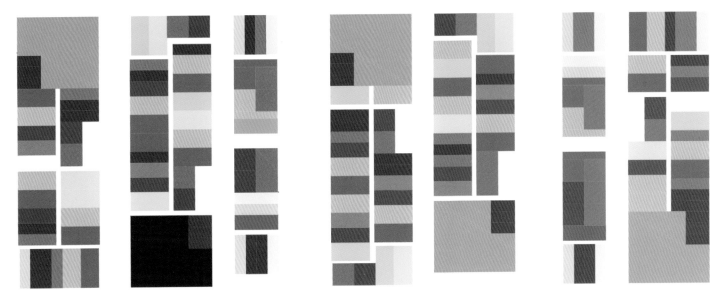

图 02 住房类型模型
Fig 02 Housing Typology Model

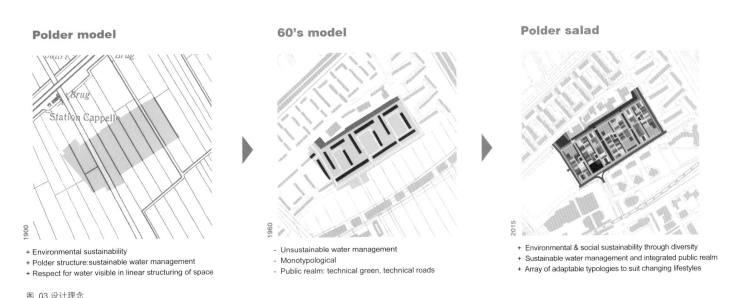

图 03 设计理念
Fig 03 Concept

欧洲青年建筑师竞赛

欧洲青年建筑师竞赛是欧洲的城市规划和建筑设计奖项，每两年评选一次。在欧洲各地，登记在案的设计场地有40余个，这些场地吸引着年轻的设计人才提出创新的设计方案来解决种种复杂的场所问题和设计难题。

欧洲青年建筑师竞赛已经举办了二十多年，它已经成为一些设计公司事业起步的跳板，这其中就不乏像MVRDV建筑设计事务所这类现已全球闻名的设计公司。"波德沙拉"是第十一届欧洲青年建筑师竞赛的获奖作品，是位于鹿特丹东部的某一场地的总体规划（图01）。

关于我们

本公司名称为"安德斯"，这个词在荷兰语中代表着"与众不同"的意思。

我们是伊莲娜（乌克兰籍）和肯·托普森（英国籍）。虽然我们都是建筑专业出身，但是我们大部分职业生涯都在从事有关荷兰的都市化以及景观设计相关领域的工作。这些从业经历使我们获得了以多学科知识背景以及超前的景观理念来进行总体规划的赞誉。我们使用的方法同以往设计中普遍运用的方法不同，以往的方法中，景观设计已经沦为了给建筑师填补空隙的角色。

Europan

Europan is a European urban design and architecture competition held once every two years.

More than 40 sites around Europe sign up to attract young design talent to think of innovative solutions for often difficult sites and complex design issues.

Europan has existed for more than 20 years and has been the springboard for the careers of now internationally renowned offices such as MVRDV. 'Polder Salad' is our winning entry for the eleventh session of Europan, a masterplan for the site, which is in the East of Rotterdam(Fig. 01).

Who we are

The name of our office is 'anders' which means 'different' in Dutch.

We, Elena (Ukrainian) and Ken Thompson (British), although having graduated as architects, have been working in the fields of urbanism and landscape in Holland for the majority of our careers.

This has given us an appreciation for a multi-disciplinary and landscape-led approach to masterplanning, a 'different' approach

图 04 理念模型
Fig 04 Concept model

荷兰景观
荷兰西部很大一部分区域都处于海平面以下，鹿特丹以及此选址均处在海平面以下约两米的位置，河流被水坝拦截，并且由一个复杂的水体管理系统承担着陆地排水的任务。电动水泵用来把雨水抽到河水中，在过去，这一工作是由荷兰标志性的风车来完成的。

荷兰都市景观类型
"圩田"是指那些通过水泵排水后的大面积土地。这些圩田被划分为一些狭窄的线性区域，宽度约为30-50m左右，被一些人工开凿的水渠分隔，用于圩田不同区域的排水（图02）。

这些水渠赋予了荷兰富有特色的条纹景观肌理。这些圩田上发展传统住房，使得现有的排水结构得以存留，形成了成行的房屋组成的城市街区，这一布局是和那些带状地区宽度的细微变化相适应的。荷兰人生活在海平面以下的窘境使得水资源管理不断创新变革，并且成为城市景观中的一种类型。阿姆斯特丹优美的运河景观和鹿特丹的标志性的风车展现了水道在作为一种工程的必要设施的同时，更能成为一道独特的景观（图03-04）。

城市化进程与荷兰田园城市
二战后的荷兰人对住房的需求急速增长，加之新的预制建筑技术和现代化进程，共同推动了荷兰主要城市周边大型郊区新城的发展。这些圩田常被视为能够开发新的房产的空白区域，与此同时，那些现状良好的水渠网络却很少受到重视，它们没有被整合到新城市框架结构中。结果，很多开发区中的绿地因常常积水而无法使用。此外，由于住房的类型在质量、多元性以及适应性方面的缺陷，很多此类开发区越发不受欢迎，开发的可行性降低。近年来，这些住房区域被重新规划，荷兰也在探求可行并且更为持续的替代方法。

波德沙拉
我们的选址就是这样一个战后的开发区，卡佩勒·安·丹·艾瑟是临近鹿特丹东部的一个新城花园。这一地域的历史发展地图展示出来以往圩田水坝的结构，这些水坝被填平用于建造战后的住房。

我们的意图在于强化圩田的结构（水渠的密度而非它们的准确位置）并且为一些住房区域创造新的景观框架，水体被整合为一系列新

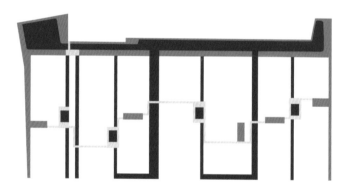

图05 用绿色蓝色标识的结构图（绿色代表绿地，蓝色代表水体）
Fig05 Green / Blue structure

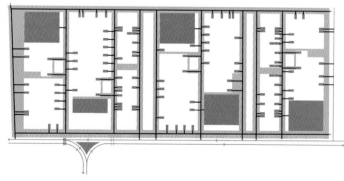

图 07 基础设施和停车场
Fig 07 Infrastructure and parking

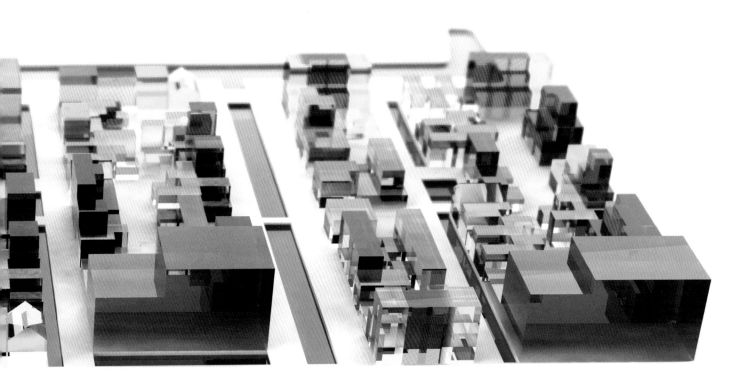

6 住宅类型
6 Housing types

的城市景观形态（图05）。这些景观架构限定出一系列人文尺度的城市街区，它们被多样化的可栖居类型所填满——也就是Salad，它们可以实现一种和现有的开发区相似的密度。这些单类型的住宅街区将被一定数量的房子所替代，这些房子同地面相连接并且有着自己的私家花园（图06）。

设计理念
都市景观

这些水道被整合为一种新的城市景观层次，其中混合了不同数量的水体层、公用的表面以及绿化区域（图07）。这些为住房区域提供了相应的框架。在它的南面，轮廓鲜明的边线提供了一个面向城市中心的布局匀称的正面景观；在它的北面，一个软质的绿色边界把现有的东西方向的水道连接起来。

通过这两个边界的连接，一定数量的线性公共区域复原了所有圩田结构中的水渠。水体和一些造景用的开花果树一起软化了景观的效果并且把自然的复杂形态带入到街景中，营造了一种和花岗岩铺装的公共表面的相互对比的效果。

住宅景观

在通常情况下，当对于房屋概况和数据的解读是抽象概括的时候，

to one which has become prevalent, where landscape has been relegated to the role of filling in the bits left over by architects.

The Dutch landscape

A large part of the West of Holland is below sea level. Rotterdam and our site are around two metres below sea level.

The river is held back by dikes, and a complex water management system drains the land.

Electric pumps are used to pump the rainwater up into the river, a job which used to be performed by the iconic Dutch windmill.

Dutch urban landscape typologies

A 'polder' is the name given to the large area of land that is drained by these pumps.

It is divided into a number of narrow linear fields of around 30-50m wide divided by hand-dug ditches, which were used to drain these polder fragments(Fig. 02).

These ditches are what give the Dutch landscape its characteristic stripes.

Traditional housing development on these polders retained the

被忽视。简化设计中的问题是建筑师的必要设计手段，但是当这一手段被用在大尺度的设计中时，可能导致平庸的、重复的和不人性化的环境的产生。经济危机表明荷兰房地产业中多度充裕的住宅已经成为一个问题。

我们在多样化和复合性方面看到了解决途径，并且也通过专注住房质量和关注人的需求来探求解决方法。这些方法已经通过发展不同的住房类型来吸引不同的生活方式和市场得以实现。它们不仅在提供更加可持续的经济模式方面有一定的优势，而且创造了一种更为人性化的环境，这是一种带有自然的复合形态的住房景观（图08–12）。

通过多样化实现的可持续性

我们的规划旨在处理战后仓促发展导致的不足，这一过程是通过专注于质量的文脉方法以及发挥水体管理的技术优势来实现景观的潜在价值的。圩田沙拉通过自身的文脉性和自身的多样性来实现可持续性，我们希望这个作品能够如同急风骤雨一般带来更加健康的都市化的回归。□

existing drainage structure, creating urban blocks of row houses which adjusted to the small variations in the widths of the stripes of the polder.

The Dutch predicament of living below sea level has led to some innovative transformations of water management into urban landscape typologies.

The beautiful canals of Amsterdam and singels of Rotterdam show how waterways can become more than just engineering necessities(Fig.03-04).

Modernism and the Dutch garden city

The urgent need for mass housing in post-war Holland, coupled with new pre-fabricated building techniques and the ideas of the modern movement led to the development of large new-town suburbs around the major cities in Holland.

图08 局部效果图
Fig08 View of Polder Salad

图 09-12 局部效果图
Fig09-12 Views of Polder Salad

The polders were often treated as a blank slate on which to develop the new housing, and little regard was paid to the existing fine network of ditches.

They were certainly not regarded as a quality that could be integrated into the new urban framework.

Consequently, many of the green areas in such developments are often water-logged and unusable.

In addition, due to a lack of quality, variety and adaptability in the housing types, many of these developments are becoming unpopular and unviable. In recent years these housing areas are being re-developed, and the Dutch are looking for ways of ensuring that what replaces them can be more sustainable.

Polder Salad

Our site is one such post-war development. Capelle aan den IJssel is a new-town garden suburb to the East of Rotterdam. A historical map of the area revealed the pre-existing structure of the polder ditches which had been filled in to create the post-war housing.

Our proposal is to reinstate the polder structure (the density of ditches rather than their exact locations) and to create from these a new landscape framework for the housing area, water being integrated into a series of new urban landscape typologies (Fig. 05).

This framework defines a series of human-scale urban blocks, which are filled-in with avariety of adaptable dwelling types – the 'salad', which achieves a similar density to the existing development.

The monotypological apartment blocks will be substituted with a substantial number of houses that have contact with the ground and their own private gardens(Fig. 06).

Design Philosophy

Urban landscape

The watercourses are integrated into a hierarchy of new urban landscape typologies which mix the layers of water, shared-surface and green in different quantities(Fig 07).

These provide the framework for the housing plots. To the South a hard edge provides a formal frontage towards the city centre. To the North a soft green edge relates to the existing East-West watercourse. Connecting these two edges ,a number of linear public spaces reinstate the ditches of the original polder structure. The water, together with ornamental flowering fruit trees soften and bring natural complexity into the street profiles, creating a contrast to the granite paving of the shared-surfaces.

A landscape of housing

Architects often forget when reading housing briefs that statistics are an abstraction – an average.

The urge to simplify design problems is an essential skill for architects, but when this is applied to larger scales it can result in banal, repetitive and inhumane environments.

The economic crisis has shown that the over-abundance of apartments on the Dutch housing market is problematic. We see a solution in diversification and hybridisation, and also by concentrating on quality and peoples' needs. This has been achieved by developing a number of different house-types to appeal to different lifestyles and markets. This not only has the advantage of providing a more sustainable economic model; it also creates a more humane environment, one with natural complexity - a landscape of housing(Fig 08-12).

Sustainability through diversity

Our plan aims to address the shortcomings of the hasty postwar development, through a contextual approach drawing on the qualities and the potential of the landscape by making a virtue of the technical problem of water management. Polder salad aims to achieve sustainability through its contextuality and its diversity. We hope that it hails the return to a healthier kind of urbanism.■

作者简介:
安德斯建筑与都市设计 / 荷兰鹿特丹

Biography:
Anders Architecture and Urbanism / Rotterdam, Netherland

1. 警衛室
2. 生態池
3. 石板橋
4. 拱橋
5. 生態小島
6. 茶亭
7. 樹屋
8. 茶席小丘
9. 榕樹休憩平台
10. 香草花園
11. 靜心水台
12. 公園入口A
13. 公園入口B
14. 公園入口C
15. 公園入口D
16. 公園入口E
17. 樟樹山坡
18. 烤肉亭
19. 陽光草坪
20. 賞花小徑

图 01 三千坪公园平面图
Fig 01 Layout of 10,000 m² (3000 pings) park

国城建设《高雄小城－一亩田》
—— 一个生态小区的规划
DESIGNED ECOLOGY
—— A DIFFERENT APPROACH TO COMMUNITY PLANNING

洪嘉聪　　　　　　　Chia-Chong Hong

项目位置：台湾高雄市路竹区
项目面积：143,892 m²
委托单位：国城建设
设计单位：国城建设设计
景观设计：国城建设设计部
完成时间：2013 年 6 月
获　　奖：2012、2013 年高雄市透天建筑景观类建筑园冶奖

Location: Kaohsiung City, Taiwan
Area: 143,892 m²
Client: Kuo Chen Development and Construction Co., Ltd.
Designer: Department of Design, Kuo Chen Development and Construction Co., Ltd.
Landscape Design: Department of Design, Kuo Chen Development and Construction Co., Ltd.
Completion: June, 2013
Awards: 2012 & 2013 Architecture and Landscape Award

图 02 小区道路种植乔木，别具四季变化风情
Fig 02 Trees along the community streets dress in different shades for the different seasons

前言

高密度的小区集体开发案，有如工厂生产线般大量制造、复制、再堆砌块状建筑，像是注入了某种漠然巨大的粘稠物质，间歇破坏土地、自然及人之间的关系，切割了彼此互动、依存的联系；渐渐地，人们开始思索并寻找自身对于住家的真正渴望与想象：想要重回有天有地、有树有水、那样纯净清明的低密度居住空间。基于这样的想法，国城建设《高雄小城－－亩田》开始对土地、对自然、对家屋、对人…有了精彩的对话与深刻的探寻。

本小区位于南台湾高雄市路竹区、距离五分钟就是中正商圈及路竹火车站，占地14公顷，放眼望去一片绿意盎然：有大岗山、田园、菜圃、蔗田、树林…，这是《高雄小城－－亩田》的建造基地，也是在台湾难得一见的大面积、可开发为住宅的土地，对于任何建筑团队而言，都是梦想与挑战的综合体。这块素地遂由半亩塘江文渊建筑师的环境规划整合，以及庄辉煌、曾永信、王东奎、陈俊廷、林泽森、洪嘉聪、郑立伦与坂井俊夫建筑师群的美学才华努力下，集结众建筑师最美最好的思维，内化成为《高雄小城》最深层的底蕴，除了具备都市难寻的绿野蓝天，更考虑到人类的群居特性，使远离尘嚣的乡村闲情与聚落共居的守望相助，均能一并兼顾（图01-02）。

景观绿化 万物生意盎然

凝聚建筑师塑造基地的核心目标：「打造一个优美、永续、生态的优质别墅小区」，为了达成这个目的，自动退缩空地、留设口袋公园及大型小区公园，散落在这个14公顷的土地上。房屋的高度不超过3楼，栋距留到最大，让每户都有最好的采光条件。《高雄小城－－亩田》总户数约为400户，主入口进来即为1公顷的生态公园。为了营造本小区多元的景观，更规划散落小区各处的口袋花园，共约1,800平方米（图03-08）。

由于本小区的规划产品都属于大地坪的别墅，因此住户的同构型较高，对于小区环境的绿意形象塑造，成了许多住户最在意的事情，甚至自动自发将前庭、后院、阳台及露台种花莳草、摆酒瓮，加上宽敞明亮的落地窗与轩窗，提供了介于室内外的日照、落雨缓冲空间，让南台湾的大把日光、悠游于天空草原之间的清新空气，任凭住户无限取用，使整个小区变得更为美丽（图09-12）。

「生态永续」小区开发新价值

国城建设对于《高雄小城－－亩田》景观绿化及生物多样性相当

图 04 入口意象－小区入口两侧种植老樟树，形成绿色隧道
Fig 04 Entrance imagery– A green tunnel formed by two rows of old camphor trees greet residents at the entrance

图 03 经由众建筑师的努力，《高雄小城－－亩田》荣获2012、2013年建筑园冶奖
Fig03 Through the efforts of a brilliant architect team, Greenfield won the 2012&2013 Architecture and Landscape Award

Introduction

High-density community development runs on the mode of factory mass production. The developers mass-replicate units and pile them up into building blocks, like planting gigantic sticky objects that are indifferent to the environment. Again and again, they destroy the land, the nature, and their relationships with man. They cut off all rapport existing between the environment and the people living within. On and on, man began to see and began to seek the home in their distant memory and imagination. People want to return to the kind of home where they can see the sky, the soil, trees and water, and this yearning translates into clean and bright low-density living. From conceptualization of Greenfield, Kuo Chen Development and Construction Co., Ltd. began a series of dialogues and profound exploration to the land, the nature, the concept of home, and the people.

Project "Greenfield", located in the Luzhu District of Kaohsiung, is five minutes from the Zhongzheng Road business circle and Luzhu Railway Station, spreading across a land of 14 hectares with a panoramic view of immense Dagangshan hills, fields, vegetable paddies, sugarcane plantations, woods, this is the site of Greenfield- a rare jewel of large residential estate development project in Taiwan. To any architecture team, development of such scale is a dream, as well as a challenge. Planning and

05 街道边退缩绿地，形塑开放式空间
05 Green buffer zones by the streets create an open space.

重视，除了公园绿地，小区内的联络道路也退缩1至5米的绿地空间，以作为公共的绿化使用。这些绿地空间都种植大型乔木，形塑生态绿廊，让小区内绿意盎然、虫鸣鸟叫。为了让小区道路更具四季风情变化，种植醉娇花、赛赤楠等本土种的小乔木。

利用生态工程学的概念，创造适合生物永续成长繁衍的栖地。以多样化的乔木、灌木、地被植物提供鸟类与昆虫栖息地与食物来源。本小区将1公顷的生态公园作为生物种源，透过小区道路两旁的绿廊串联各区，再加上散落小区的口袋公园，足以打造一个生态小区所需要的栖地空间，让本小区内的生物更具多样性。

生态湖公园设计则仿造自然栖地，营造小型生态系统：湖域降低微环境气温，以湿生、挺水、沈水、浮叶四种水生植物，自然吸引水鸟、青蛙、萤火虫等生物。而多层次林木结构：大、小乔木、灌木、地被植物，可提供鸟类筑巢所需材料：树枝、草、叶子、树洞助其栖居。水池中央也设有一个生态小岛，种植高密度的乔木、灌木。加上水池隔离人为干扰，让小动物可以安心栖息繁衍（图13-16）。

多样化地貌 增添绿光家园风采

规划《高雄小城－一亩田》的公共设施空间，有别于都市住宅大楼，以提供户外活动空间为主题，包括生态池、树屋、水池边的静心平台、茶席空间、有机菜园、阳光草皮等……，提供住户大空间休憩娱乐场地，多样化的地景面貌使得乡居生活不再平淡乏味，并增添了生活乐趣与人情暖意。除了主要的三千坪水岸绿地外，更在每个小区之间口袋公园提供近距离的小区互动空间，几乎是从自家后院就能直接走入另一方流水潺潺的芳草碧连天之间，一层又一层的缠绵绿意，不仅只是人工造景，并结合了电缆、弱电系统地下化的贴心设计与主动式绿建筑

design of this virgin land was commissioned to the team of Ban Mu Tang led by Architect Chiang Wen-Yuan for holistic planning and a team of architects, Chuang Hui-Huang, Tseng Yong-Hsin, Wang Tung-Kui, Chen Chun-Ting, Lin Che-Sen, Hong Chia-Chong, Chen Li-Lun, and Sakai Toshio. This dream team of architects transcends their best imaginations for the concept of home into the soul of Greenfield. This is a city tribe with an azure sky above a community designed for the human population away from the hassle-bustle of the city (Fig01-02).

Green landscape nurtures a vibrant ecology

We set a goal to build a quality residential estate integrating the elements of aesthetics, sustainability, and ecology. To achieve this goal, the designers planned sufficient empty spaces for maximum comfort and a series of pocket parks and community parks throughout the 14 hectares of land. Every house is designed no taller than three stories with maximum spacing between houses for optimum penetration of natural light. Greenfield is community of 400 households. A one-hectare ecological park opens the vision right after the main entrance and pocket parks with an accumulated area 1,800 m^2 scatter across the estate, creating a landscape of varied visual sensations (Fig03-08).

This project features large units. The environment is an integral part of their lifestyle; therefore, some residents decorate their front/back yards, porches, and balconies with beautiful flowers, green

图 06 生态池公园，赋予别墅小区永续生态的新生命
Fig 06 Eco-pond park brings new life to the sustainable ecology of the community

图 07 漫步公园绿荫林间步道，使人神舒意畅
Fig 07 A stroll along the walkways shaded by lush greens refreshes your mind

图 08 错落于各区的口袋公园，将自然元素引入居住环境
Fig 08 Scattered pocket parks extend elements of nature into the living spaces

09 将前庭后院景观犹如公园纳入自宅，尽享阳光绿意大自然
09 Immerse the living spaces in the relaxing views of the yard and cozy air under the sun

图 10 在自家庭院悬挂吊床，享受绿意清新舒适感
Fig10 Hang a hammock and enjoy refreshing air in your own backyard

图 11 露台种花莳草增添生活情趣
Fig 11 Have fun showing off your green thumb from your balcony

12 透过窗从室内飘向街道，阳光草叶任悠游
12 Sunlight and shadows of trees and leaves take an excursion from the windows to the streets

图 13 仿造自然栖地营造小型生态系统，万物得以生生不息
Fig 13 The small ecosystem echoing nature creates a habitat for the regeneration of lives

图 14 生态池中央小岛种植大树，提供生物栖息繁衍
Fig 14 Large trees planted on the isle at the center of the ecological pond provide a safe habitat for animal reproduction

图 15 以水生植物降低湖域气温，吸引自然生物栖息
Fig 15 Aquatic plants lower the temperature around the lake to attract diversified habitats

作品实录 PROJECTS

16 小型生态池成了绿头鸭驻足栖息地，成群结队悠游又自在
16 The small ecological pond has attracted a flock of wild ducks

17 建造孩童梦想的树屋，提供住户休憩娱乐之所
17 The dream tree house is a fun house for the children

图 18 体验农夫有机种植，随心所欲亲近大自然
Fig 18 Organic vegetable garden brings the residents direct contact with the nature

的环保节能,让绿色环保思维深植小城,齐心为自然生态尽一份心力(图17-18)。

《高雄小城－一亩田》主要诉求为低密度的别墅小区,留设大面积的绿地空间,除了各户的前庭后院,对于小区公共的绿化也着力甚深。行道树主要都采用可以形成绿色隧道的大型常绿乔木,如樟树、土肉桂、杜英、台湾榉,栽种间距都达 6 米以上,且种植在自然土壤,让乔木得以自然生长。更值得一提的是,《高雄小城－一亩田》所采用的樟树为本基地的原生树种,在此生长已达 40 年以上,选择在合适季节(11 至 2 月),断根移植至合适位置,让这些老树得以继续生长茁壮(图 19)。

期望《高雄小城－一亩田》住户看到更多触手可及的天蓝水碧草青,感受大地如何涵养水分,阳光则转化水蒸气负离子,抚育这个愿意与自然和平共处的小区,而不再有高楼大厦遮蔽的清爽徐风、带走只属于拥挤城市的燥热湿溽。让每个住户的身心,都能调节到最舒缓自然的状况,同时也因为理解了自然的美好,共同编织更多绿意与笑语并存的小城故事。□

plants and decorative things, like ceramic crafts. French windows and view windows provide sufficient sunset and ventilation, and the passionate sunlight of southern Taiwan and fresh air roaming in between the sky and green fields provided by nature in abundance turns the estate into a beautiful community (Fig09-12).

Sustainable ecology creates a new value for community development

As a developer, Kuo Cheng Real Estate Co. placed its focus on the landscape and biodiversity. Therefore, in addition to the parks, all access roads leading to the community are planned with one to five meters of green buffer zones planted with large trees. Rows of large evergreen trees create a green corridor for development of a healthy ecology where green leaves sprout vibrantly and bugs and birds chirp enthusiastically. The community roads are embraced by smaller shrubs, like scarlet bushes and Acuminate Acmena to create varied views for the different seasons .

Through ecological engineering technology, we set out to create a habitat for sustain ecological development. The diversified trees, shrubs, and ground covering plants provide habitats and food sources for birds and insects, and the one-hectare ecological park is the center of biodiversity. The green corridors link the pocket parks to create an optimum habitat for a highly diversified ecosystem.

The eco-lake park is designed to mimic a natural habitat for development of a small ecosystem. The water in the lake lowers the temperature in the micro-environment, and the wetland, emergent, submerging, and floating plants attract water birds, frogs, and fireflies. The stratified vegetation structure is formed by the large and small trees, shrubs, and ground covering plants. The branches, grass, leaves, and tree cavities provide suitable environment for nesting birds. The ecology isle at the center of the pond is planted with high-density trees and shrubs, and is isolated from human interference to create a safe habitat for small animals (Fig13-16).

Varied landscape creates pleasing visual sensations

Planning of the public spaces in Greenfield is different from condominium buildings in the city. The focus is placed on the spaces for outdoor activities. A wide range of facilities, such as the ecological pond, tree houses, meditation podiums, tea pergola, organic vegetable garden, and immense lawns, provide residents spaces for outdoor recreational activities, and the varied landscape replaces the dull and colorless cityscape with fun and interesting touches of humanity. Beyond the nearly 10,000 m2 (3000 pings) of watery shores and greenery, each section of the community is serviced by a pocket park, linking the backyard of each household to the green landscape and water in the open field. Layer after layer of greenery was created from not only landscaping technology but also attentive design of underground power cable and transformer systems and energy-saving facilities. The environmental-friendly town was designed with the concept of protecting the environment in mind (Fig17-18).

"Greenfield", designed as a low-density housing estate, features large green spaces. Every unit enjoys independent front and back yards, and holistic community environment planning provides the residents lush greens and fresh air. Large evergreen trees, such as Camphor, Indigenous Cinnamon, Common

图 19 移植原生树种，沿续生生不息的绿光家园
Fig 19 Native plants shade the green homes for generations to come

Elaeocarpus, and Taiwanese Zelkova, were chosen as boulevard trees. Each tree is spaced over 6 meters apart and planted in natural earth for proper branching and healthy development. One thing worth noting is that the Camphor trees planted in "Greenfield" are native plants, which have been growing in the local environment for over 40 years. They were transplanted to suitable locations during the preferred season (November to February) to ensure high survival rate (Fig19).

For Greenfield, we set out to create a housing estate where the residents are surrounded by azure skies, green water, and lush greenery and feel the land nourishing vibrant lives, the sun transcends moisture into negative ion-charged phytoncide, and nature nurtures a community embracing its endowment in harmony. This is a community without high-rise concrete structures blocking the cheering breezes or the hot and human air exclusive to the crowded cities. Residents in the community live in an environment where they can relax and return to the original state of harmony, and the goodness of nature is the reason they choose to create their own stories for generations to come. ■

作者简介：

洪嘉聪／男／国立成功大学建筑研究所／国城建设建筑师／台湾高雄市

Biography:

Hong Chia-Chong / Male / MA, Graduate School of National Cheng Kung University / Architect of Kuo Cheng Development and Construction Co., Ltd./ Kaohsiung, Taiwan

QUATTRO 住区景观
RESIDENTIAL LANDSCAPE OF QUATTRO

宝克·科布东赛迪　　Pok kobkongsanti

项目位置：泰国曼谷
项目面积：8,022.97m²
委托单位：Sansiri 房地产公司
设计单位：TROP 景观设计有限公司
设计总监：宝克·科布东赛迪
项目设计：丛凡·阿迪彩
设计团队：安维特·奇瓦它蓬，那塔蓬·瑞泰，
　　　　　蔡查奂·班吉斯瑞
完成时间：2011 年
获　　奖：2012 美国景观设计师协会住宅设计类荣誉奖

Location: Bangkok, Thailand
Area: 8,022.97m²
Client: Sansiri PLC
Design Organization: TROP Co.Ltd (TROP : terrains + open space)
Design Director: Pok Kobkongsanti
Project Designer: Chonfun Atichat
Project Team: Anuwit Cheewarattanaporn, Nattapong Raktai, and Chatchawan Banjongsiri
Compeltion: 2011
Award: Honor Awards, Residential Landscape design, American Society of Landscape Architects (ASLA), 2012

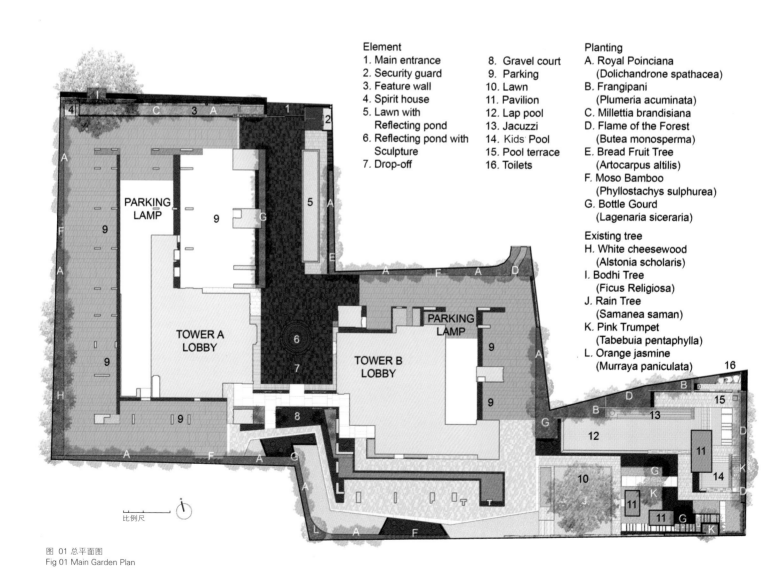

图 01 总平面图
Fig 01 Main Garden Plan

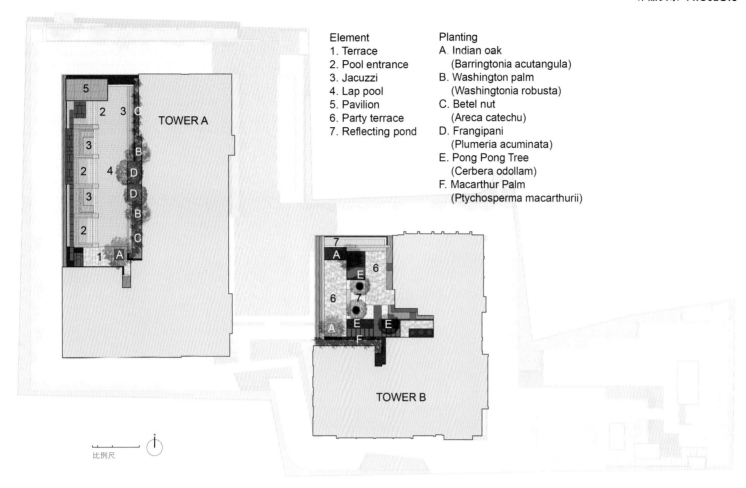

图 02 屋顶花园平面图
Fig 02 Roof Garden Plan

Project Statement

Three gardens designed for a new Condominium in the middle of over-crowded Bangkok(Fig 01-02). The main idea is to respect the existing "Residents", which are old giant Rain trees (*Leguminosae mimosoideae*) and their small inhabitants (squirrels and birds) and to encourage the new residents to live with Nature. The Rain Trees become the heart of our design, while all other garden elements are created to complement those trees(Fig 03-04).

Project Narrative

Quattro is a High-end Residential Project in an up-scale area of Bangkok. Look at Bangkok now. You can hardly find soil in a city so full of concrete and asphalt. Here people have changed their ways of living too. In the past, we may have preferred to live in a small house, with a small garden, outside the city areas, and commute in and out of the city daily. Not anymore. To fit the present day's fast paced life style, It is quite normal these days that a younger generation of Thais are moving to the city and choosing to live in high-rise buildings, instead of their out of town homes. Horizontal Living is out. Vertical living is the thing to do.

Basically now we live in a box. Stacked boxes, to be precise. Ranging from 30 sqm to larger ones, they are still boxes. These concrete boxes stacked on top of each other take the place of soil. This is far from the ideal "Home" that everyone has in his/her mind. If you ask kids to draw their ideal "Home", you would surely see similarities between each drawing. It may have a house, a garden of some sort, a fishpond, etc. Compare that to the condominium boxes we are living in, we may still have a house, but the garden and the pond are gone.

Well, this is not good, and we would like to change it.

项目说明

这三个花园位于人口稠密的泰国曼谷城市中心区，是为新的共管式公寓而设计。（图01-02）该设计的主要构思在于尊重"原住民"的存在，包括一些大的雨豆树（*Leguminosae Mimosoideae*）和这些树上的小"居民"（松鼠和鸟），并且鼓励新的居民与自然共存。这些雨豆树成了花园设计的核心要素，而其它的造园要素都是为这些树木锦上添花（图03-04）。

项目概述

Quattro是位于曼谷繁华地段的一个高档居住区项目。如今的曼谷，在满是钢筋水泥的城市中已经很难见到土壤。在这里居住的人们也已经改变了他们的生活方式。过去，人们可能宁愿生活在城郊一栋有着小花园的小房子中，每天通勤去城里上班。但是，现在不同了。为了适应当今快节奏的生活方式，年轻一代的泰国人搬进城市，离开他们镇子里的家，并选择住进高层建筑中。眼下，这已经是再平常不过的事了。贴近地面的生活已成过去，住在高层楼房里才是当下潮流。

我们现在好像生活在一个盒子里。更为准确地说，好像是生活在堆叠起来的盒子里。无论它们是30平方米或者更大，都是些盒子。混凝土的房子一个个地堆叠在彼此的上方，而不是在土地上。这与每个人头脑中"家"的概念相去甚远。如果你让孩子们画出头脑中的"家"，一定会看到很多共同点。这些家有座房子和某种样式的花园，还有个小鱼池等。如果把这种家的形象和我们的住宅相比，我可能还有一个房子，但是花园和水池都不复存在。

那么，让我们来改变这件不太好的事情吧。

图 03 面向高层 B 的主体亭子看到的景观，一个特别设计的石制表面也被融入到这个
半闭的空间中
Fig 03 View from the main pavilion looking towards Tower B. A special designed stone facade is added to create a semi-enclosed space

图 04 从上面俯瞰下去，这个主花园好像同那些原有的树木一样，从一开始就存在于此
Fig 04 Looking from above, the main garden appears unchanged from its previous state with the original trees

图 05 主花园水池别具匠心地设计远离树木根部，水池高于地面且没有边界，用来营造"纯净之声"。原生树木得以原地保留
Fig 05 The main pool is strategically placed away from the existing trees' root balls. The pool is raised, with an infinity edge, to create 'white noise'. We keep the trees exactly as they were before

6 所有的元素都是从矩形演变而来的
6 Every element is created from rectangular shapes

图 07 在一些特定的点，镜子被加入景墙中，使得整个园子看起来更宽敞
Fig 07 At certain points, mirrors are added into the featured wall which malce the gard looks like bigger

主花园

Quattro 这一项目的原址是一栋有着大花园的老房子。当我们第一次到该场地时，看见了那些在场地中生长了 40 多年漂亮的雨豆树。我们脑中第一个念头就是："要不惜一切代价把这些树保留下来！"。在业主、建筑师和其它法律顾问的帮助下，我们成功地把两个高层建筑物的选址从这些树旁边挪走了。这么做，虽然让业主失去了很多可出售的地皮，却让那些能够体现场地特征的珍贵树木得以存留。

下一步，就是在这些树旁边布置主花园。我们计划在曼谷的中心建设一座都市绿洲，通过研究这些树木和它们对周围环境的影响开始设计，如阳光被树木细小的叶片所遮蔽，从而可营造树下温度宜人的绿茵空间。树干下的空间无疑是放置新的休憩凉亭的好位置。为了最大程度地减小对于树木根球的影响，我们尽量减少使用大的硬质铺装，而是通过一系列安置于不同位置的露台来替代。通过软景、鼠尾草类植物（珍珠草）和他们之间的卵石，营造了局部的小气候。主体水池的选址在离树根较远的地方放置，通过布设高于整个花园四周大约 80cm 高的边墙，这种无边界的设计和外溢的水流创造了"纯净之声"的效果，在降低外部交通噪音方面起到了很大的作用（图 05-06）。

设计智慧，这一灵感来自于项目的名称（4）Quattro。所以我们运用了一系列矩形的组合，把它们与三维空间的园林构图相组合。一些大的矩形成为了露台，与此同时一些小的矩形成为了步道。一系列的矩形的石头同样被用在亭子的墙面上，从而为使用者创造了半封闭的空间（图 07-08）。

Main Garden

The site of Quattro was an old house with a big garden. When we first visited the site, we saw beautiful rain trees, living in the middle of the site for more than 40 years. Our first thought was that "we have to save those trees. No matter what it takes." Working closely with the Owner, the architects and other legal consultants, we managed to locate 2 tall residential towers around the Trees. By doing so, the Owner lost much of their selling space, but, instead, we can keep the price-less feature of the site, the Trees.

Our next move is to locate the main garden around the trees. Here we planned to create an urban oasis in the middle of Bangkok. We started by studying the trees and the effect they made on the area around them. For example, Sunlight is basically filtered through their tiny leaves, creating comfortable temperatures for the space below. Of course the area below their big branches become our new seating Pavilions. In order to minimize any impact on the Trees' root balls, we avoided large hardscape areas. Instead, a series of terraces are added here and there, with softscape mouse tail plants (*Phyllanthus myrtifolius* Moon), and small pebbles in between, creating a microclimate effect. The main pool is also located away

wall, and raised at approximately 80cm from the garden, the Infinity Edge and its over-flowing water create a "White Noise" effect, which also greatly helps reduce the outside traffic noise (Fig 05-06).

In terms of the design, the inspiration comes from the project's name, Quattro (4). So we play with a combination of rectangular shapes, combining them into 3 dimensional garden compositions. Some bigger rectangles become terraces, while the smaller ones are used as stepping pathways. A series of rectangular stones are also used on the pavilions' walls, creating semi-enclosed spaces for their users (Fig 07-08).

Overall, the Main Garden is a memento of our new residents' old out-of-town homes—a big open space with nothing in between the ground and the sky, except the Trees' huge canopies—a rarity for this kind of project.

图 08 主花园里保留下来原生美丽的雨豆树
Fig 08 View of the main garden. 2 beautiful existing rain trees and other trees are kept in the new design

总体来说，主花园的设计是新住户对于原来郊区家园的纪念，除了树木围合成的巨大穹顶以外，还造了一片空旷的开敞空间，这在此类项目中十分罕见。

两个屋顶花园

在高层A，一个台层样式的水池被别具匠心的布置在停车场的顶部。这个水池不仅帮助了住宅单元的销售，而且水池也为低层的住宅单元降低了热量。由于屋顶的面积很小，没有办法留出更多的绿地空间，而用水池旁边种植低矮的绿篱来替代。通过选择了两棵特别的鸡蛋花树（图09），其高度可以超过水面，树影同样帮助减少了周围的住宅建筑之间的反光。

高层B同主花园相距不远，因此在这里不需要水池。我们在此创造了能够容纳不同活动的露台广场，运用类似降低热量的方法，将露台分成两块小的硬质区域，中间用一个小的反射水池分隔（图10-12）。

露台上种植了曼谷市树——红花玉蕊（*Barringtonia acutangula*）和百花海芒果（*Cerbera odollam Gaertn*），有利于改善局部的小气候（图13-14）。

尽管新居民是年轻一代的泰国人，但亚洲的生存文化与西方家庭不同一个家庭里三世同堂很常见，在Quattro住区里也是一样。一些人把自己的父母带来，一些人马上也要成立家庭。所以，在这里我们想要为他们创造的不仅是一个花园，而是一个特别的场所。

这是一个可以颐养天年的地方，老人可在这里享受舒适的空间，居民可在此养育子女。这是一个为三代人而建的的、名副其实的都市绿洲（图15-16）。

图 09 从高层A看到的水池露台的景观，特别选择了一些枝条延展到水中的鸡蛋花
Fig 09 View of the Pool terrace on Tower A. In a tight space, special Frangipani trees were selected which have branches extending over the pool

图 10 露台被分成很多小部分来减小热量和眩光，中间加入了水和卵石
Fig 10 The terrace is divided into smaller areas, to reduce heat and glare, water and pebbles are added in between

图 11 为了最大化的实用空间，尽可能地减小了硬质露台的面积，同时为阳光浴的椅子设计了树荫池，两边设有按摩浴缸
Fig11 To maximize the use of space. Hard scape terrace is keep to minimum. Instead a shallow pool was proposed for lounge chairs, with Jacuzzis on the side

图 12 高层B的露台广场，这一构思使得这个园子好像是建在地面上而不是建在停车区域上
Fig12 Tower B's party terrace. The idea is to make the garden appear like it is built on the ground, instead of on the parking structure

图 13 在此应用乡土树种百花海芒果和红花玉蕊，它们不需要太多的养护管理
Fig 13 Local trees, the Pong Pong Tree (*Cerbera odollam Gaertn.*) and the Indian oak (*Barringtonia acutangula*) are used here because they require low maintenance. (*Cerbera odollam Gaertn.*) (*Barringtonia acutangula*)

图 14 水池的反射被用来帮助屋顶散热，一些独特设计的种植细节展现了低调的外貌，而不是那种夺人眼球的炫目设计
Fig 14 A reflecting pond is added to cool down the roof. Special planter details are designed to give a modest appearance, instead of a "look at me" design

图 15 小巧灵活的屋顶布局能够满足日常活动
Fig15 Small but flexible, the roof layout can be adjusted to fit the day's activities

图16 在一天中的某个特定时刻，叶片的阴影为墙面增添了独特的情趣
Fig16 At certain times of day, shadows from the leaves create beautiful patterns on the wall

Two small gardens are also created on each of the towers

At Tower A, a Pool Terrace is strategically located on top of its parking structure. This pool not only helps to sell the units, its water also greatly helps reduce heat for the lower floor units.

Because of the roof's small size, we do not have much green area left. Instead of creating small strips of plantings around the pool, we handpicked 2 special Frangipani trees (Fig 09), that extending over the water. Their shade also helps reduce the glare that will be reflected into the surrounding units.

Tower B is close to the Main Garden, so we do not need another pool here. Instead, we created a Party Terrace for multipurpose activities. Using a similar idea to reduce heat, the Terrace is divided into 2 smaller hardscape areas, with a reflecting pond between them (Fig10-12).

Two of Bangkok's local trees, the Indian oak (*Barringtonia acutangula*) and the Pong Pong Tree (*Cerbera odollam Gaertn.*), are planted here to create a microclimate effect for the terrace (Fig13-14).

As a result, both terraces can be used all day, despite Bangkok's infamous heat.

Even though our new residents are the younger generation of Thais, the Asian way of living is still different from that in the West where it is quite usual to have 3 generations of family members living in one household. At Quattro, it is the same. Some residents have already brought in their parents, and some will have their own family soon. What we try to create for them here is not just any garden, but a special place.

This is a place where people can spend the rest of their lives; a place where the elderly have a comfortable outdoor space; a place where one can raise children; a true urban oasis for all 3 generations (Fig15-16).■

作者简介：
　　宝克·科布东赛迪／男／景观设计师（设计主管）／TROP 景观规划工作室／泰国，曼谷
Biography:
　　Pok kobkongsanti / Male / Landscape Architect (Design Director) / TROP Co.Ltd / Bangkok, Thailand

查尔斯·沙（校订）
English reviewed by Charles Sands

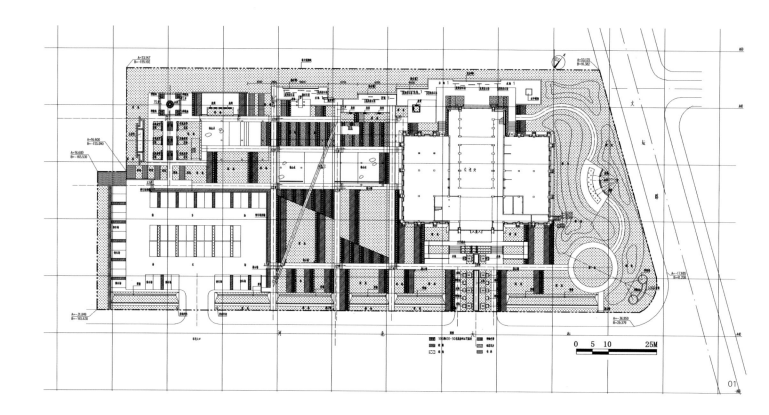

理性与感性的交融
秦皇岛远洋海悦公馆展示中心景观设计
FLOW OF THE ORDER
LANDSCAPE DESIGN FOR THE DISPLAY CENTER OF YUANYANG HAIYUE RESIDENTIAL AREA, QINHUANGDAO

杨珂   Ke Yang

项目位置：秦皇岛市海港区西白塔岭村
项目面积：13,000m²
委托单位：秦皇岛市海洋置业房地产开发有限公司
设计单位：R-land 北京源树景观规划设计事务所
方案设计：章俊华、白祖华、李晶、杨珂、陈一心
    天野真（WAS 设计事务所）
扩初＋施工设计：章俊华、胡海波、杨珂、丁玲、夏强、
    曹可为
专项设计：朱彤（结构）、杨春明（电气）、白晓燕（建筑）
施工单位：上海园林（集团）有限公司北京分部
完成时间：2010 年 8 月

Location: Bai taling Village, Harbour district, Qin huangdao City
Area: 13,000m²
Client: Qin Huangdao Ocean real estate investment
Landscape Design: R-land Beijing Ltd
Designers: Junhua Zhang, Zuhua Bai, Jing Li, Ke Yang, Yixin Chen, Makoto Amano (WAS)
Devepment and Construction Design: Junhua Zhang, Haibo Hu, Ke Yang, Ling Ding, Qiang Xia, Kewei Cao
Special design: Tong Zhu (structure design), Chunming Yang (electricity), Xiaoyan Bai (architecture)
Construction: Shanghai Landscape (group) co., LTD., Beijing Branch
Completion: August, 2010

图 01 平面图
Fig01 Master plan

注：本文图片由章俊华和 R-land 北京源树景观规划设计事务所提供
Photography: Junhua Zhang and R-land Beijing Ltd

作品实录 PROJECTS

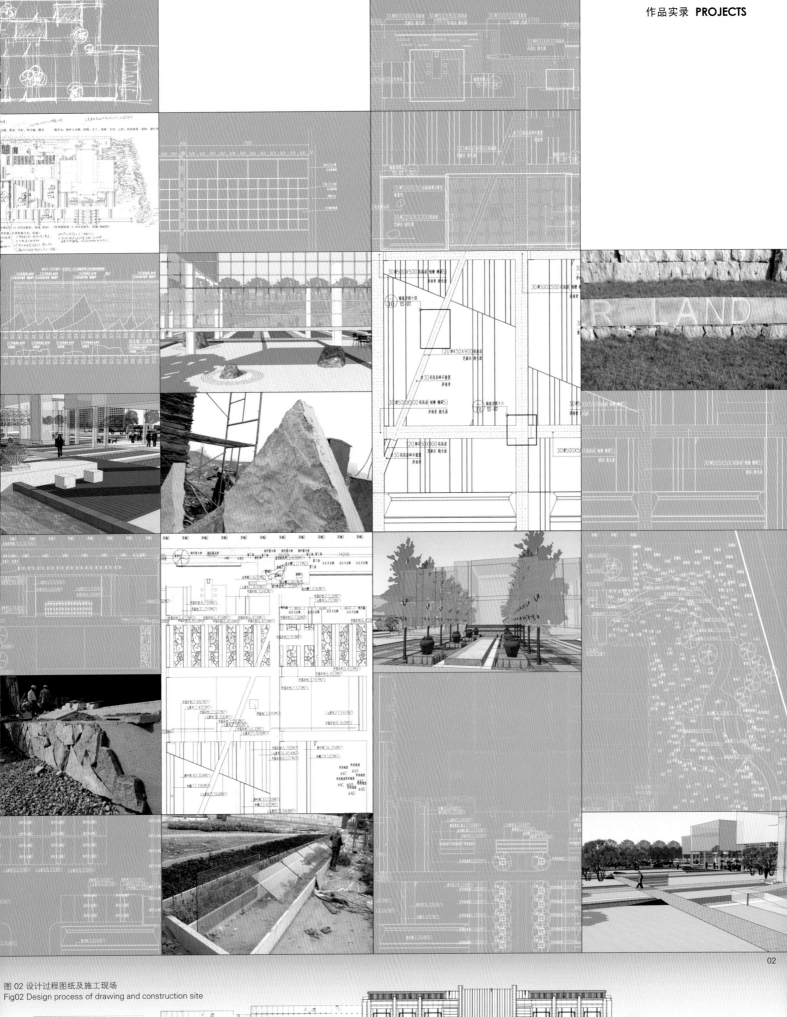

图 02 设计过程图纸及施工现场
Fig02 Design process of drawing and construction site

简欧风格的展示中心建筑（图01-03），层高近12m的大厅，气派、端庄。西侧由3个架空的样板间玻璃方盒子将地块分隔为前广场（图04-05）、后花园及东区3个主要的户外空间，最西端是一个面积比前广场还略大的停车场（图06）。由于场地东高西低（高差1.5m），为此将停车场周边统一抬高，并在前广场设置3处台阶，使得停车场在高差上有别其它空间的同时，与前广场、后花园（图07）、（图08）东区一起形成4块风格相异，但又不可分割的相互关联的空间场所。最终由棋盘式的网状路径将场地有机的联系在一起（图09）。

前广场空间通过相互平行交错的3条带状分隔带的重复出现，表达抽象的"海浪"印象，并选植一棵高8米冠幅7米的特大白皮松作为此场所在空间体量上的平衡（图08）。同时利用一条宽1米的斜线（路）连接前广场与后花园。规整、纯粹、极简、超然是前广场空间的特质表现。后花园空间分部延续了前广场序列的线形，在细部的刻画上通过多层次地被植物的自然种植（图07）、流水墙高矮、长短、前后的错落及水面凹凸、进退、宽窄的收放，表现后花园无唯不尽的小尺度变化。西端的礼仪空间及儿童小游园的场所设置，满足了甲方对项目的多重要求。丰富、优雅、日常、人性是后花园空间的特质表现。东区采用当今流行的微地形起伏＋复层密植的"常规"自然设计手法，表现"城在林中"的场所意境（图10-11）。平和、凝静、野趣、亲切是本空间的特质表现。停车场作为前广场空间的背景，成排成行的树列反衬前广场开敞的空间，周边自然面的毛石墙及10cm厚的压顶，表现素朴、厚重又略显奢华的场地氛围（图06）。即表达安稳流动的视觉感观，又不失欲透而不露的空间特质。

斜纹拼花的路面铺装，架空层下海兰色电光釉材质的饰面，2:1高宽比的页岩挡墙，海滨习语镶嵌的条石，梁下悬空的雪格屏风（图09），主入口处对置的饰品坐椅与花钵（图04），碎石中散置的卧石，展示大厅对景的溪溪叠水及波浪浮雕，大规格自然石墙的海螺装饰，分隔带石条板的附视效果及理性表现（图12），碎石勾边的斜坡草坪，错落有致的水杉林，错台式种植及成排的序列乔木，丛生的西府海棠与白桦（图13），当地野生地被节节草的应用……均希望表达出设计师不经意中的刻意追求——理性与感性的交融。□

图03 背光处的水池，勾画起人们对"慢"生活的向往
Fig03 Pool in the backlight

图04 清晨中略显朦胧的入口空间，散溢着礼仪中的品质
Fig04 Entrance space in the early morning

图05 光影散落在石纹间，感受场所的宁静与超然
Fig05 Light scattered between the stone texture

图06 微地形与丛生白桦，装点着停车空间的外沿
Fig06 The outer limits of the parking space decorated with micro topography and *Betula platyphylla*

图07 小尺度的地被种植，让来访者憧憬日常的美好生活
Fig07 Exquisite ground covers

The architecture of the Jane European Style Exhibition Center (Fig.01-03) is a 12m-high grand hall. To the west is the main exterior space, which is divided into a front plaza(Fig.04-05), a back yard garden and the east area by three square open-floored glass rooms with a parking lot a little larger than the plaza to the far west (Fig.06). As the site is 1.5m higher on the west side, the designers decided to raise the surroundings of the parking lot and lay steps in three different places in the front plaza. Thus the comparatively lower parking lot connects with the front plaza, the back yard (Fig.07)garden and the east area (Fig.08) through a network of paths (Fig.09).

图08 绿篱、园路、碎石、玻璃挡板、彩叶草等软、硬线条的简捷表现
Fig08 Simplification of the garden elements

图09 寓意海浪的线形与海兰色电光釉面砖,以"形"与"色"抒发对地域的情怀
Fig09 Lines related to waves and the glazed tiles

图10 朝阳中的绿地、园路映照着清馨的天空与白云
Fig10 The grass and path in the morning

图11 背光中的树影与芒草、血草、狼尾草…交织成美丽的自然画卷
Fig11 Shadow of trees and ornamental grasses

图12 夜色中的机刨面分隔条仍保持着线形的空间肌理与建筑的立面的光点遥相呼应
Fig12 The divided strips which surfaces carved by machine in the night

图13 透过海棠枝叶的间隙，编织着永无静止的画卷
Fig13 The view seen from branches

作者简介：
杨珂 / 男 / 工作室主任 / R-land 北京源树景观规划设计事务所 / 中国北京
Biography:
Ke Yang / Male / Dean / R-land Beijing Ltd / Beijing,China

查尔斯·沙（校订）
English reviewed by Charles Sands

The front plaza expresses abstract "waves" through the undulation of three integrated parallel green strips. A huge lacebark pine, 8m in height and with a 7m canopy, is planted to balance the space (Fig.08). An oblique road links the front plaza to the backyard garden. The road, which is orderly, pure, simple and transcendent, contrasts to the backyard garden, which is colorful, graceful, common and humane. This allows one to experience the liner sequence between spaces, expressed in the detail of

13

the multi-tiered plantings and the irregular flow of water from the water walls (Fig.07). Large-scale baroque spaces and a children's playground to the west children's playground to the west accommodate the diverse needs of the residents. The peaceful, quiet, and sincere east area applies the method of micro-topographic variation and dense, tiered plantings to create the artistic conception of "City in the Forest" (Fig.10-11). As the background of the front plaza, the parking lot embodies an atmosphere of simplicity, dignity and a touch of luxury with its rowed tresses (Fig.06), and rough walls surrounding a 10cm raised roof. This creates a blurring of edges and an uninterrupted visual continuity.

A wide range of complementary and contrasting textures and colors converse in features such as thetwill-patterned road, the ultramarine blue glazed material under the open-floor (Fig.09), the shale wall witha 2:1 height to width ratio, the white screen hung under the beams, the flag stones combined with scattered stones, the wave-like relief facing the hall (Fig.04), and the sloping lawn rimmed with scattered stones. The planting of dawn redwoods(Metasequoia glyptostroboides), midget crabapples (Malus micromalus), white birch(Betula platyphylla)(Fig.13), and branched scouring rush(Equisetum ramosissimum) All contribute to the intuitive, over-arching theme of the designers -- a flowing order.■

大一山庄住宅景观设计
DAYI VILLA COMMUNITY

广州高雅房地产开发有限公司
Guangzhou Gaoya Estate Development Co. Ltd.

项目位置：广州市白云大道北
项目面积：200,000m²
建筑面积：100,000m²
委托单位：广州高雅房地产开发有限公司
设计单位：广州高雅房地产开发有限公司
景观设计：马里奥·博塔、马丁·罗班、麦克·亨特、温立新、郑时龄、梁志天、项秉仁、赵红红、马岩松等
完成时间：2009年05月
获　　奖：2012年园冶杯住宅景观奖工程金奖，亚洲十大超级豪宅、中国十大超级豪宅榜首

Location: Baiyun street north, Guangzhou, China
Area: 200,000m²
Building area: 100,000m²
Client: Guangzhou Gaoya Estate Development Co. Ltd.
Designer: Guangzhou Gaoya Estate Development Co. Ltd.
Landscape Design: Mario Botta, Martin Robain, Michael Hund, WenLixin, Zheng Shilling, Liang Zhitian, Xiang Bingren, Zhao Honghong, Ma Yansong
Completion: May, 2009
Awards: 2012 Yuanye Engineering Gold Cup Award for Residential Landscape, Asia's top ten super mansion, China's top ten super mansion

作品实录 PROJECTS

图01 大一山庄
Fig01 Dayi residential community

图02 绿树掩映下的大一山庄
Fig02 Forestry Dayi villa community

项目介绍
项目定位

针对金字塔顶尖的人群不仅仅是单纯对居住面积的简单追求，并不是单纯追求极尽豪华的装修，而是逐步去寻求更高层次的精神满足。需要一个远离喧嚣、洗尽凡心的筑梦环境，享受悠然的自然景色，孕育广阔的心灵空间，在宁静祥和中自然释放自我，而大一山庄（图01-02）正是为了满足此类顶尖人群的需求而造（图03-05）。在理念上，大一山庄除了是建筑的博览园，更应是人居的艺术大境，遵循"颂扬千年的儒释道思想"与"现代建筑文明"的完美融合，力臻将"豪宅"和"艺术"合二为一，用自然、朴素、和谐的环境去创造人们的生活所求，打造一个近乎世界级的、超乎人们意料的建筑艺术博览园，这是一个充满"人性化、个性化，艺术性、诗情画意般"的超级大师手笔。（图06-07）

为确保大一山庄理念的完美实现，我们聘请到寰球顶尖建筑大师和优秀的建筑师团队，在近20万平方米的土地上塑造仅百余栋艺术群雕。这些大师来自于世界各地，有当今的建筑泰斗——瑞士建筑大师马里奥·博塔，世界第三大建筑事务所——美国NBBJ建筑事务所，世博会总规划师、法国政府公认的建筑大师——法国AS建筑工作室马丁·罗班，中国科学院院士——郑时龄，中国首位建筑学博士——项秉仁，中国房地产十佳建筑影响力设计师——赵红红，当今最具影响力的20名设计师之一——马岩松等等。

在大一山庄每一栋都是建筑大师投入精力和智力的杰作，每一件作品均力图高度的精致和特别，以求让创作者深厚的文化内涵得以淋漓尽致的表现，带给人们一种身份象征和一种美学取向价值。每一栋

Introduction
The ideas of the project

For those who live in the top of the pyramid of material life, what they really need is not just a residential area or deluxe decoration, but rather a higher-level fulfillment of the spirit—a place far from the noisy world that can clarify the mind. In Daya Villa, people can enjoy beautiful scenery and foster a space for their soul. They can find themselves in nature and meet all of their needs (Fig 01-02). Dayi villa community is an exhibition of buildings, and at the same time an artistic dwelling place for people(Fig 03-05). It is a perfect combination of "one thousand years of Daoism" and "modern civilization" in architecture. The ideas of "mansion" and "art" are combined together, using natural, original and harmonious methods to create what people want. A world-class garden of architecture is created—a masterpiece full of humanity, character and poetry(Fig 06-07).

To make the idea of Dayi become a perfect work, we invited top architects and A class design groups to make the buildings become true works of art in an area of 200,000m². These architects come from all around the world, among them there are some big names in this field, such as Mario Botta from Switzerland; NBBJ one of the world biggest architectural firms; Matain Robain from AS architectural firm, the designer of the general planning

作品实录 PROJECTS

悠然的自然景色
Beautiful natural scenery

与自然亲密无间
Close to nature

贴近大自然
Neighbouring nature

诗情画意
Like living in a painting

美不胜收
Fantastic

图 08 夜幕中的大一山庄
Fig 08 Dayi at night

图09 人与自然和谐相处
Fig09 Harmony of men and nature

图10 精致如艺术品
Fig10 Looking like an artistic

1 山间绿树掩映着鲜花
1 Flowers blooming among trees

图 12 绿水长流
Fig12 Still water runs deep

图 13 林中小亭
Fig13 The pavilion

都像稀世艺术品一样，令创作者、居住者都从内心深处感受到强烈的视觉震撼（图08-10）。

绿色低碳，环保节能

大一山庄除了努力打造每栋建筑的艺术性及唯一性，在社区规划中，"大一"充分尊重自然湖泊、坡地山丘，保留原生坡地地貌。仅北区就保留及全冠移植了超过2000棵百年原生树木，如国家一类保护植物活化石桫椤、世界上唯一开蓝色花朵的蓝花楹、顶级香水原料的白兰、珍惜古沉香等数千棵珍贵树种，通过有序的景观布置，形成四季各异，花香不断的景观效果，并形成一套完整的生态链条，俨然一座森林公园，在炎热的夏天，园区内空气温度比市区要低 1 - 2 度。除此之外，不惜增加了 5000 余万的成本，反复修改项目入口园林的规划，只为保留了园林入口处的百年古榕（图 11-18）。

在建筑设计上，大一山庄还广泛采取绿色低碳材料以及科技手段，如充分利用了文丘里效应，最大化地利用自然风，形成室内的自然风循环系统，让室内的热气向上被室外的自然低温吸走，从而使室内保持凉爽；绝大多数建筑设置屋顶花园，外立面也多采用浅色石材；每栋建筑使用了 LOW - E 中空防辐射镀膜玻璃幕墙以及百叶窗等，起到隔热降温的作用，从而减少室内的能耗；而采光天井及下沉式庭院的广泛应用，也是为了起到地下室自然采光以及通风的作用；以引水库循环活水用于花园灌溉、水系打造，使用高效低能耗建材等等，都旨在引导一种自然生活方式，让人们更加的贴近大自然，回到人的自然属性。

房流于林影，人行于画中

在园林景观设计方面，融合白云山的自然生态环境体系，集合现代造园手法与古典园林韵味，在现代性的基础上创造大气、自然、和谐、

of the Shanghai EXPO and a well known architect in France; Zheng Shiling, an academic from the Chinese Academy of Sciences; Xiang Bingren the first doctor of architecture in china; Zhao Honghong, one of the top ten most influential architects in the real estate industry; and Ma Yansong, one of the 20 most influential contemporary architects.

Every building in Dayi is a masterpiece made from hard work and intelligence. What you can see here is a kind of high-class elegance and specialty. Each work is a perfect presentation of the culture of the designer, and also a demonstration of social identity and artistic taste. They are all rare artworks. Both the designers and the residents can be deeply moved by the views (Fig 08-10).

Low carbon, environment-friendly community

In its community planning, Dayi tries to keep the natural appearance of the original lakes, sloping fields and native slopes. 2000 original trees were kept in the north zone, such as the cyathea——a kind of plant known as a living fossil—which is under the highest class of plant protection; the only Jacaranda with blue blooms; prynne-which contains a material used in perfume; ancient angerwood; and many other kinds of precious trees. These tree species are arranged to produce seasonally blooming scenery, at the same time an integral ecological link is formed. As in a forest park, the temperature in the summertime is 1-2 degrees lower here. As well, to protect the hundred-year old banyan tree,

14 山庄一角
14 A corner of the community

图 15 生态房屋
Fig15 An ecological house

图 16 植被茂盛
Fig16 Plants flourish

图 17 水影婆娑，美景如画
Fig17 Beautiful like a painting

图 18 植被种类繁多
Fig18 A variety of plants

图 20 水流潺潺
Fig20 Flowing water

图 21 流水瀑布
Fig21 The waterfall

图 22 人在画中游
Fig22 Like walking in a painting

图 23 世外桃源
Fig24 Land of idyllic beauty and peace

图 24 山中一隅
Fig24 A corner of the hill

图 25 山间步道
Fig25 Pathway in the hill

具有中国特色的人间奇景。"房流于林影，人行于画中"是本项目园林规划设计的理念。通过自然、随意、现代又具古典园林意境的设计风格结合山地高低变化丰富、植被自然生态的特点，使大一山庄成为"虽由人作，宛如天开"、巧夺天工的艺术精品，达到"人间仙境"的境界。

大一山庄信奉古人所云："天生万物以养人，人取天之美养其身"。所以对于园林设计的要旨是：进入大一山庄的园林，使人徜徉于艺术文化的绿色空间中，精神得到慰藉，心情放松，有所牵挂也忘却了牵挂，有浑然忘机，真正实现天人合一（图19-22）。园林小中有洞天、世外有桃源，一石一叶都有情，用滴水之意造就音乐之章，都是力求打造最自然的动、静之态，而绝无人工造作之痕（图23-25）。

专属定制，顶级配套

在顶级配套方面，为匹配项目的顶级定位，斥2亿元巨资打造由国际级新锐设计师马岩松设计的奢华会所，它是一个专为精英人群营造一个罕见的"七星级贵族享受"平台。

聘请了加拿大高力国际提供一流的物业服务，以"世界顶尖酒店式物业管理"彰显居者非凡的身份。样板房智能化方面选用美国快思聪系统（微软比尔·盖茨未来之屋选用品牌），通过指尖轻触操纵，妙触未来智能家居生活。

再加上户户私家电梯、中央空调系统、中央吸尘系统、地热系统、供水净水系统、光纤到户、有线电视、中国电信等专属定制配置，可根据客户需求不同而选择定制。

the planning of the entrance was revised adding about 50 million yuan to the project cost (Fig 11-18).

In the design of the buildings, low carbon, technical methods such as the Venturi effect have been applied to make full use of natural wind currents so the interior hot air can be drawn out by the exterior low-temperature air through the integral air circulation system. Most of the buildings have roof gardens, and vertical surfaces are clad in light colored stone materials. Every structure is designed with LOW-E radiation coated hollow glass curtain walls and shutters, lowering the temperature and reducing energy consumption. Lightly colored patios and sinking yards are also widely used to help bring natural lighting and ventilation into the basement.

Buildings in wooded shadows, people in paintings

The natural system of the Baiyun mountain, is a combination of modern construction methods and the charm of classical gardens. The concept of this project is "Buildings in the shadows of woods and people in picturesque scenery", it uses a natural, casual, modern style with classical characteristics, a combination of varying topography and ecological features, to produce "a man-made work of natural art".

There is a saying in China that embodies the philosophy of Dayi villa, "people benefit from things born in nature, and take advantage of the beauty of nature to cultivate themselves". So the main point of the design is——when you enter the garden of Dayi, you can feel a green space with culture and art, it is a relaxing palace where you can forget your troubles and discover the harmony between heaven and people (Fig19-22). In this garden, you can find an expansive world and a hidden paradise. The stones and the leaves are all imbued with affection. The water creates music to accompany the dynamic and static patterns of nature without a trace of human pretensions (Fig 23-25).

Exclusive customization, top facilities

In order to match the project's high standards, 200 million RMB were budgeted to build a world-class cutting-edge luxury clubhouse designed by Ma Yansong, in which the elite can enjoy a "royal seven-star recreation" facility.

Colliers International hired a Canadian A class property services company to ensure an extraordinary identity. U.S. Crestron systems (associated with Microsoft's Bill Gates) was selected as the model home automation system, whereby each resident has complete control over all aspects of their intelligent home.

What is more, each household contains a private elevator, central air conditioning system, central vacuum system, geothermal systems, water purification systems, FTTH, cable TV, and other exclusive custom configurations that can be customized to each resident's different needs. ■

作者简介：
广州高雅房地产开发有限公司 / 房地产开发与建设 / 中国广州
Biography:
Guangzhou Gaoya Estate Development Co,. Ltd. / Real Estate developing and constructing / Guangzhou, China

查尔斯·沙（校订）
English reviewed by Charles Sands

图 01 布洛涅·比扬古公园
Fig 01 The Parc de Billancourt

专题文章 ARTICLES

法国布洛涅 – 比扬古公园
PARC DE BILLANCOURT, BOULOGNE-BILLANCOURT, FRANCE

亨利·巴瓦　米歇尔·欧斯莱　奥利维耶·菲利浦　　Henri Bava　Michel Hoessler　Olivier Philippe

　　布洛涅－比扬古公园（图01）以自身的位置和规模构成了塞纳河新区的景观基石。布洛涅－比扬古公园的形状来自新建区域开发项目整体的主要轮廓，这一区域位于北面一个大面积的近梯形城市街区和毗邻塞纳河畔的城市街区之间。源自沿塞纳河岸的线状外形和下凹式的高程设计，使得布洛涅－比扬古公园呈现出了类似于沿河码头式的空间结构。布洛涅－比扬古社区公园虽然被城市中林立的楼群所包围，但当地居民无需进入其中，便可享受其作为一个城市绿肺（图02）所带来的益处。布洛涅－比扬古公园是一道景观屏障，混凝土台阶便于游客席地而坐，尽情享受公园的乐趣。同时，公园会随着季节的更替和塞纳河的水位变化而呈现出不同的景观。

Because of its size and location, the Parc de Billancourt (Fig.01) constitutes the landscape cornerstone of the new Rives de Seine district. It takes its shape from the main outline of the overall project for the new district located between two entities: the large city blocks of the north "Trapèze" and the blocks adjacent to the Banks of the Seine. Its linear shape and hollowed heights surrounded by banks gives the park the appearance of a planted dock. The garden, a green lung (Fig.02) surrounded by housing and office buildings, can be enjoyed by local residents without needing to enter it. Serving as railings, the concrete steps allow

这片面向塞纳河的空地，为布洛涅-比扬古市和塞纳河重新建立联系提供了契机。法国岱禾景观设计事务所的方案是想建立一座"城市中点缀的岛屿"，并融入整个城市规划的框架。一方面，目的是重现一个19世纪风格的景观公园，人们可以置身其中观赏风景；另一方面，也为城市的相关自然因素增添了可变性和不确定性。公园中的一部分持久不变，其余的部分随自然而改变，水位高低的变化导致了公园出现景观上的变化，从而使得公园也经历持续不断有规律的结构变化。

图 02 公园虽被众多建筑物包围，但是植被覆盖茂密，是城市的"绿肺"
Fig 02 The garden, a green lung surrounded by housing and office buildings, can be enjoyed by local residents without needing to enter it.

图 03 公园中的水池
Fig 03 The pond

图 04 水体景观
Fig 04 Water landscape

图 05 夜幕下的比扬古公园
Fig 05 The Parc de Billancourt at night

　　由于水流的频率和流量不能控制，那么就需要管理水体的分流。花园中不同层级的建造就是考虑到了水流的持久性和不稳定性。每一层都有特定的底层和植被来适应调节不同程度的水位。于是，公园有了应对水流变化的环境机制，小岛承担了积攒雨水，分流过量河水的功能。两个充满水的小水池（图03），分别在公园的两端起到调节水位的作用。

　　水体景观（图04）和与之相呼应的碎石河床，小岛，还有湿地，它们共同构成的自然环境是被废弃的支流适应流向不确定水流的结果，同时，斜梯式的河堤形似一个码头，多层级的码头满足了应对暴风雨

users to sit to enjoy the park and its changing appearance with the season and the level of the Seine.

　　This clearing, which opens onto the Seine, represents an opportunity to re-establish the link between the city of Boulogne-Billancourt and the river. Agence Ter's proposal is to create an "island of nature moored in the city", forming part of the overall urban project. On the one hand, the aim is to recreate a 19th century landscape park in which landscapes are there to be viewed, and on the other hand to inject a degree of uncertainty

图 06 夜间照明
Fig 06 Illumination at night

以及辅助泄洪等多种需要。花园集城市中心建筑群和公共区域的雨水收集和过滤功能于一身，它的设计考虑的是连贯运作而不是水位高低。

公园除了可变性的这一特征之外，它的另一个称奇之处是没有采用照明设施。出于经济和生态方面的考虑，公园的四周采取照明（图05-06），被照亮的主干道可以在夜幕降临时起到指引游人的作用，但是只有在河堤的两边才有彩色灯光。因此，这个以水为主体的公园是在游客的来来往往之间被唤醒和进入梦乡。□

and variation linked to the natural elements. Certain parts are durable whereas others, reversible by nature, undergo continual transformations resulting from changes in the water level, thus regularly redefining the garden's configuration.

Whilst the frequency and quantity of water cannot be controlled, its distribution is managed. Several levels have been developed in the park as a function of the permanency or instability of the water. Each of these has a specific substrate and vegetal panel suited to the level's degree of flooding. The park thus offers an environment that freely accepts unpredictable transitions, the extent of which can be seen in the continuity of the Seine. Islands form and change as a function of rainfall and river swelling. Two ponds(Fig.03), which are continuously supplied with water, are foreseen on each side of the park to verify the water level.

A landscape of water (Fig.04) and therefore of gravel beds, islands and marshes: this natural environment is that of an abandoned river branch subject to the unpredictability of the water flow, while the terraced banks resemble a dock with as many levels as required for the different functions (storm basin, river-swell basin...). The garden integrates the collection and filtration/oil decanting function for rainwater runoff from the central city blocks and public areas. It is designed to operate coherently regardless of water level.

Over and above its changeable character, another surprising aspect of the park is the absence of lighting. For economic and ecological reasons, the surroundings are illuminated, and the main path is marked to orient the last ramblers at nightfall (Fig.05-06), but only the constructed borders emit coloured light. Thus, this garden shaped by water goes to sleep and awakens with its visitors.■

作者简介：
亨利·巴瓦 / 景观与城市规划设计师 / 岱禾景观设计事务所 / 法国巴黎
米歇尔·欧斯莱 / 景观与城市规划设计师 / 岱禾景观设计事务所 / 法国巴黎
奥利维耶·菲利浦 / 景观与城市规划设计师 / 岱禾景观设计事务所 / 法国巴黎

Biography:
Henri Bava / Landscape and urbanisme Architect / Agence TER Paysagiste Urbanisstes / Paris France
Michel Hoessler / Landscape and urbanisme Architect / Agence TER Paysagiste Urbanisstes / Paris France
Olivier Philippe/ Landscape and urbanisme Architect / Agence TER Paysagiste Urbanisstes / Paris France

滨海一号景观绿化工程
GREENING AND ENGINEERING PROJECT IN BINHAI NO.1

范美军　王彬彬　卢云慧　祁永　　Meijun Fan　Binbin Wang　Yunhui Lu　Yong Qi

项目位置：天津滨海黄港生态开发区内
项目面积：12.18万 m²
委托单位：天津滨海黄港实业有限公司
施工单位：天津绿茵景观工程有限公司
完成时间：2011年9月
获　　奖：2012年园冶杯住宅景观奖工程金奖

Location: Tianjin, China
Area: 12,180,000m²
Client: Tianjin Binhai Huanggang Co., Ltd
Construction: Greenery Landscape Engineering Co., Ltd, Tianjin
Completion: September, 2011
Awards: 2012 Yuanye Engineering Gold Cup Award for Residential Landscape

01 滨海一号温泉度假酒店景观绿化工程全景图
01 Overall view of the project

02 滨海一号温泉酒店景观绿化工程
02 Greening and engineering project of Costal Number One Hot Springs &Spa Holiday Inn

图 03 滨海一号温泉度假酒店景观绿化工程入口
Fig 03 Entrance of the project

工程概况

滨海一号温泉度假酒店景观绿化工程坐落于天津滨海黄港开发区京津高速北塘收费站北侧（图01-03），是一家集住宿、餐饮、会议、温泉、康体休闲、健康养生于一体的商务度假酒店。酒店占地面积26.6万 m²，其中景观面积约21.5万 m²，并拥有7.5万 m²的湖面，工程总造价为1,0046,5146元。主要工程内容为土方工程、排盐工程、给排水工程、绿化工程和园林景观工程。其中，土方工程包括平整场地、地形构筑、挖湖堆山等；排盐工程包括铺设排盐管、铺设淋水层、砌筑检查井等；给排水工程包括给水管铺设、砌筑给水阀门井等；绿化工程包括乔木、花灌木、绿篱、地被、花卉、水生植物的栽植和草坪铺种；园林景观工程包括亭、廊、榭、舫、桥、园路、铺装、喷泉等（图04）。

该项目是典型古典园林的继承，在造园手法上将古典园林的四大特征："本于自然，高于自然；建筑美与自然美的融揉；诗画的情趣；意境的蕴涵"应用到筑山、理水、植物、建筑等方面，而在景观设计上重点突出"师法自然、分隔空间、融于自然，园林建筑、顺应自然、树木花卉、表现自然"几个特点，在有限的空间里营造出丰富的园林景观。

项目特点

该项目主要通过筑山、理水、植物、建筑、书画五方面来营造一个"有山有水、亭台廊阁、水榭石坊、花木虫鱼"山明水秀，富有诗情画意的景观环境（图05）。

筑山

滨海一号温泉度假酒店院内假山是主要景观之一（图06），多处掇石堆山筑于亭楼之间，看楼看山，处处有景，步移景异。

以分隔空间，丰富空间多样性并增加趣味性。南湖景观采用了中国古典园林中"一池三山"的造园手法，在湖中分别堆砌大小两个岛，使山水融为一体。

理水

在理水方面主要通过"掩""隔""破"三种方法来丰富水岸景观。
（1）掩，通过突显建筑和植物将曲折的池岸加以掩映。除主要厅堂前的平台，湖周围多处建有亭、廊、榭，该类建筑皆采用前部架空的

Introduction to the engineering project

Our greening project of the Coastal Number One Hot Springs & Spa Holiday Inn (Fig 01-03) is located at Binhaihuanggang development district in Tianjin, north of Beitang toll gate on the Beijing-to-Tianjin Expressway, satisfying its customers with accommodation, dinner service, meeting halls, leisure activities, and spa experiences. It covers an area of 266000m², of which the landscaped area counts for 215000m² including a lake covering an area of 75000m². The total project cost is 100,465,146 RMB. The landscape construction was divided into separate 5 engineering projects: the dig & cut project including the flattening of site, the construction of topography, man-made lakes and artificial rocks; the salt-draining project including the laying of salt-draining pipes; the construction of the water layer and the building of inspection wells; the gain & drain of the water project including the laying of water pipes and the constructing the gain valves; the greening project including the planting of trees, hedges, flowers and water plants and the laying of lawns; and the landscape engineering to build pavilions, lobbies, fountains (Fig 04), etc.

The engineering project inherits typical characteristics of classical gardens. The four main features of classical garden are applied, which are based on nature and arguably more beautiful than nature: the combination of architecture art and natural beauty; the appreciation of poems and paintings; the implication of artistic imagery; and for the practical use of mountain construction, the construction of water scenery with plantings. When it comes to landscaping, the project focuses on producing a variety of sceneries in a limited space.

Features of the engineering project

The landscape is a natural artistic environment containing water and mountains; pavilions and lobbies; trees and flowers; the construction of water scenery; planted areas; architectural

方法临水而建（图07），立于水面之上，水犹似自其下流出，用以打破岸边的视线局限（图08）。另外，水岸边建筑旁又有植物做装点来柔化建筑线条并与湖面景观相互映衬，使风景与建筑巧妙的融糅到一起，实现"动静结合"从总体到局部都包含着浓郁的诗情画意。

（2）隔，现场通过在湖面架小桥、筑堤、建廊、铺设汀步等手段（图09），来增加景深和空间层次，使水面有幽深之感。正如计成在《园冶》中所说，"疏水若为无尽，断处通桥"。

（3）破，由于院内湖面面积不大，因此湖岸以乱石为岸（图10），怪石纵横，错落有致，并植配以细竹野藤、朱鱼翠藻，虽是一洼水池，也令人似有深邃山野风致的审美感觉。

植物

滨海一号度假酒店内不仅局限于建筑、山水，此外更重要的是与自然植物景观的结合，植物景观主要是自然式设计，着意表现自然美。对花木的选择标准，一讲姿美，树冠的形态、树枝的疏密曲直、树皮的质感、树叶的形状，都追求自然优美；二讲色美，树叶、树干、花都要求有各种自然的色彩美，如红色的枫叶，青翠的竹叶、白色广玉兰、紫色的紫薇；三讲味香，要求自然淡雅和清幽。酒店内的植物以植物多样性为原则，以乡土树种为主，根据不同植物不同的形态特征和生理习性合理的进行配置，并很好的考虑了不同植物的季相变化，整体达到了四季常有绿，月月有花香。

建筑

园内主体建筑采用古典式风格，建筑类型丰富，有置地灵活的亭子、曲折迂回的长廊、植物蔓延的花架、小巧精致的桥，藉水而建的水榭石舫等，造型轻巧、玲珑优美、组合形式多样、富于变化。

此外，本项目具有审美文化、民族文化及艺术文化等特点。如：山水为蓝本，由曲折之水、错落之山、迂回之径等提现了审美文化。其建筑物的端庄、含蓄、雅致园林的叠山理水等体现出了我国的民族文化（图11）。园林将封闭和空间的相结合，使山、池、房屋、假山、叠水的设置排布，有开有合，互相穿插，在缀以四季应景花木，以增加景区见间的联系和景区的层次，达到移步换景的效果，给人一种"山重水复，柳暗花明"的感觉，充分体现了一种艺术文化。□

construction; and calligraphy. (Fig 05)

Rockery engineering

Artificial rocks (Fig 06) are one of the major landscapes elements of the Holiday Inn. The scenery changes as you move through it.

South Lake landscaping has adopted the method of classical garden making known as "one pond with three hills surrounding", it means to build two islands in the heart of the lake, making the water integrates with the rocks.

The construction of water scenery

Three methods have been adopted to enrich the waterfront landscape, which are hiding, separating and scattering.

First, 'hiding' means to shade the irregular bank of the pond with architecture and plants. Many open-floored pavilions and lobbies are built along the lake (Fig 07), from which it seems the water emanates. This creates the expansion of the space (Fig 08). Besides, plants serve as decorations on the side of the architecture and lakeshore connecting natural scenery and man-made architecture to be poetically appreciated.

Second, 'separating' is to bridge the lake (Fig.09). In this way, the landscape depth and range of levels is increased. As Jicheng—the most celebrated master of Chinese classical gardens in the Ming Dynasty—said in Yuanye, an ancient book on landscaping, "If the waterscape is unlimited, bridge it."

Third is to 'scatter' the water. As the lake is relatively small, deformed rocks are irregularly set along the banks (Fig 10) with small bamboo species and little known rattan planted within them. People may feel like they are living in the wild when they see fish swimming and green algae floating in these tiny pools.

图 04 有限的空间营造丰富的景观
Fig 04 Limited space with enriched landscapes

05

Plants

The plants highlight the holiday inn as they are in harmony with the architecture. Three standards guarantee the best choice of plants: natural beauty of the outline, a variety of colors and pleasant smells. Firstly, high standards should be met in terms of the tree crown, branches, barks, and leaves; Secondly, a variety of colors should be achieved by planting different things ranging from maples and bamboos to mangnolia and crape myrtle; Thirdly, a preference is given to fragrant plants. Various kinds of plants make it possible that different plants can be appreciated with the change of seasons.

Architecture

Major works of architecture in the holiday inn are adopted from the classical style. There are also various types of landscape features, including pavilions, wind galleries, flower blooming pergolas and exquisite bridges.

What's more, this project reflects aesthetic, ethnic and artistic cultures. The aesthetic culture is embodied by the interactions of rocks and waters; the ethnic culture (Fig 11) by containing much but revealing little of the architecture; and the artistic culture by the combination of closed and open spaces, and leveled sceneries the change with your step.■

作者简介：

范美军／男／森林资源与休憩／内蒙古农业大学学士／天津绿茵工程有限公司副总经理
王彬彬／男／作物栽培与耕作学／中国农业大学博士／天津绿茵景观工程有限公司
卢云慧／女／草业科学／内蒙古农业大学硕士／天津绿茵景观工程有限公司董事长
祁　永／男／草业科学／中国农业大学博士／天津绿茵景观工程有限公司总经理

Biography:

Meijun Fan / Male / Forest resources and preservation / Bachelor of Inner Mongolia Agricultural University / Vice General Manager of Greenary Landscape Engineering Co., Ltd, Tianjin

Binbin Wang / Male / Crop cultivation and farming system / PhD of China Agricultural University / Manager of Technology Department of Greenary Landscape Engineering Co., Ltd, Tianjin

Yunhui Lu / Female / Pratacultural science / Master of Inner Mongolia Agricultural University / President of Greenary Landscape Engineering Co., Ltd, Tianjin

Yong Qi / Male / Pratacultural science / PhD of China Agricultural University / General Manager of Greenary Landscape Engineering Co., Ltd, Tianjin

查尔斯·沙（校订）
English reviewed by Charles Sands

图 05 诗情画意
Fig 05 The scenery looks like a poetic painting
图 06 假山堆叠
Fig 06 Artificial rocks
图 07 前部架空的廊桥
Fig 07 Openfloored gallery
图 08 湖旁多建有亭台楼阁，水仿佛从其下流出
Fig 08 Water appearing to emanate from the pavilion
图 09 湖水以乱石为岸，无尽的山林野趣
Fig 09 Deformed rocks irregularly scattered along the lakeshore, guiding its visitors into the wild
图 10 湖面架设的廊桥
Fig 10 The bridge across the lake
图 11 端庄、含蓄的建筑风格，彰显民族特色
Fig 11 Ethnic culture is fully embodied

无庶小区景观规划设计
LANDSCAPE DESIGN OF WUSHU RESIDENTIAL COMMUNITY

江苏大千设计院有限公司
Jiangsu Daqian Design Institute Co.,Ltd.

项目位置：江苏省南京市玄武区锁金村9号
项目面积：13,271.16 m²
委托单位：南京长发房地产开发有限责任公司
设计单位：江苏大千设计院有限公司
景观设计：李晓军、汪辉、俞涛
完成时间：2011年
奖　　项：2012 园冶杯住宅景观设计金奖

Location: Nanjing, Jiangsu Province
Area: 13,271.16 m²
Client: Changfa Real Estate Co., Ltd
Designer: Jiangsu Daqian Design Institute Co.,Ltd.
Landscape Design: Xiaojun Li, Hui Wang, Tao Yu
Completion: 2011
Award: 2012 Yuanye Designing Gold Cup Award for Residential Landscape

图 01　小区内古典氛围的环境
Fig 01　An environment with classical style
图 02　修竹间的凉亭
Fig 02　Pavilion surrounded by bamboo
图 03　小桥流水，修竹叠石
Fig 03　Bridge, water, bamboo and stones
图 04　亭台楼阁在水面上映照出倒影
Fig 04　The shadow of the buildings
图 05　小区中的圆形拱门
Fig 05　Arched doorway
图 06　修竹掩映的步行小径
Fig 06　Pathways

图 07 洞上
Fig.07 Archway

图 08 不规则的路缘石看似随意,却是精心布置
Fig 08 Stones along the road appear strewn at random, but are meticulously arranged

图 09 古典闲境
Fig 09 Classical design

图 10 植物和拱门组成的多层次的空间效果
Fig 10 A multi-faceted space created by plants and archways

项目背景

无庶小区，位于南京市玄武区锁金村9号。地处玄武湖畔，临近紫金山，与情侣园及规划中的玄武湖北扩商业街区隔路相望。二类居住用地，占地面积13271.16 m²，容积率≤1.3，建筑面积≤18000m²，绿地率30.3%（图01-05）。

文化内涵

无庶，即"没有众多"，稀缺之意。"无"在中国文化中具有深远而广泛的哲学意义，代表了一种物质的最初形态，无即有，有即无，以无代有。用于案名，传达出项目传统文化风格，同时贴合国人审美情趣与高尚情怀。有无问题是传统玄学文化的中心问题，玄学家们认为，道德规范、礼法教化等是为"有"，天地万物的本真状态和人类的自然本性就是"无"。（图06-12）

人文院落无庶的规划正是基于这一理念，试图通过住宅院落的建筑设计宣扬传统礼法教化，强调"人因宅而立，宅因人得存"，将主人的品行、心性、涵养、意趣与宅相通。内修之时同样不能忽视本真天然之气，筑园而居，与自然最大化接触，苍木拙石保持最初模样。居者心安，即为吾乡（图13-16）。

设计说明
总体规划理念

无庶居住区以再现自然山水为设计的基本原则，追求建筑和自然的和谐，达到"天人合一"的效果。但并不是简单地模仿，而是"本于自然，高于自然"，把人工美和自然美巧妙地结合起来（图17-

Background

Wushu Residential Community is located in Xuanwu District, Nanjing, close to Mount Zijin and along the bank of Lake Xuanwu (Fig01-05).. It covers an area of 13271.16 m² with the construction area of more than 18000m² and green space accounting for 30.3% of the site.

Cultural Relations

The term "*Wushu*" has a deeply philosophical meaning in Chinese. "Wu" is the initial state of nothingness—nonexistence. But nonexistence is itself, existence. To replace existence with non-existence is of paradoxical importance. The name of the project shows this traditional cultural idea of China and approaches its essence. Being the focus of Chinese traditional Metaphysics, the paradoxical relationship between *nothingness* and *being* has been well considered. Metaphysicians think that morals and social principles are part of existence, while the initial state of everything on earth and human nature is considered as non-existence (Fig06-12).

The planning of this community is based on the above ideology, in an attempt to spread traditional cultural principles through community construction. The interaction between people and their homes is stressed, relating the dweller's character and interests to his house. To perfect oneself, one cannot go far from

图 11 墙角的一丛竹　　图 13 卵石小径与石桥
Fig 11 Bamboo in a corner　Fig 13 Pebble footpath and stone bridge
图 12 庭中的一棵树　　图 14 亭台水榭让人仿佛置身古典园林中
Fig 12 Giant tree in the yard　Fig14 The classical buildings of traditional Chinese gardens create a historical feeling

18）。这一思想在造园当中的具体表现就是"因地制宜"，善于利用现有的自然环境条件，体现出人工建造对自然的尊重与利用。"虚实相间，以虚为主"，强调建筑群体之间的关系。

（1）结合项目的先天条件和楼盘的开发思路，依据纯中式与现代相结合的新中式造园理念，营造出"素雅宅第，灵秀江南"的金陵风情（图 19-21）。

（2）在设计上采用徽派清白简素的淡雅色调，园中叠山理水，建亭筑台，莳花种树，通过移步换景、以小见大、以实化虚等手法，再现自然造化之神奇。

（3）建筑坐北朝南，户型方正仁和，设置前庭后院，在形、神上达到现代人居宅养心养性的意旨。

（4）"园中院"布局，独立入宅通道，下沉式庭院，开放式景观步道，空间更加合理，居住感闲适（图 22-23）。

nature. Living in a garden and getting close to nature is what one has to do. Home is where one feels at home (Fig13-16).

Introduction of the project

Design concept

The basic designing principle is to recreate hills and water in nature. That doesn't mean to simply imitate, but rather to combine natural beauty with artificial art in pursuit of harmony between nature and construction, so as to achieve the unity of heaven and man(Fig17-18).

(1) A simple but elegant residential landscape with Nanjing characteristics is created by combining the existing geography with the design concept, and following the ideas of neo-Chinese garden making that is the combination of traditional and modern

墙壁上的雕花
Flower carving on the wall

practices(Fig19-21)

(2) Light colors in the Anhui Style are adopted, and through rock-pilling, water features, pavilions, trees and flowers, and various garden making methods, the mystery of nature is drawn out.

(3) The square constructions are located to the north and face south, with frontcourts and a backyards, achieving the goal of pleasing inhabitants physically and mentally.

(4) The spatial logic can be expressed as "a garden within a garden" with an independent pathway oriented to the door, a 'sink' courtyard, and an open landscaped footpath. This rational spatial distribution enables the residents to live more comfortably(Fig22-23).

(5) When designing, the natural geography is taken as the basis for the entire planning process. It is the central focus of the plan.

Design principles

(1) To focus on key features, to arrange elements as simply as possible, and to follow the traditional Anhui style(Fig24-26)

(2) To adjust the design to local condition (Fig27-30)

(3) To maximize the effect of green space

从拱门看到的景观
The scenery from the arched doorway

图 17 树丛的影子和古典的亭子共同营造了静谧的氛围
Fig 17 The shadow of the trees and classical pavilion create a quiet atmosphere

图 18 拱门与门前的石灯的形状相协调
Fig 18 Stone lamps coordinate with the doorways

图 19 雕花和盆景的结合
Fig19 The stone flower-carvings

图 20 漏窗兼具审美和实用的双重功能
Fig 20 The dual functions of the decorative latticework window

图 21 由景墙组成的错落空间
Fig 21 The space made by wall

图 22 两株红枫在白墙前
Fig 22 Two maples along a wall

图 23 角落中的小池
Fig 23 A tiny pool in the corner

图 24 盆景、假山和植物的搭配
Fig 24 Rockery and plants

图 25 透过拱门，看见远处的小径
Fig 25 View from the arch of the footpath

图 26 仿佛不是身处闹市，而是古典园林中的一角
Fig 26 You may feel as if you were in the corner of a classical garden rather than in a city

图 27 小径和一排修竹尽显幽静的氛围
Fig 27 Trails and a row of tall bamboo filling the quiet atmosphere

图 28 廊的两侧一面是闭合的，一面是开敞的
Fig 28 Side of the gallery

图 29 绿植丛中的一条小径
Fig 29 A small road passing through the plants

图 30 园中一角
Fig 30 One corner of the garden

图 31 铺地和置石的结合
Fig 31 Pavement and the stones

图 32 古典元素的运用上体现了较为统一和谐的效果
Fig 32 Harmonious elements

图 33 透过拱门看到的小径通向幽静之处
Fig 33 A quite place beyond the archway

图 34 方形门
Fig 34 Square archway

图 35 花窗
Fig 35 Decorative latticework windows

图 36 花形门
Fig 36 Flower-shape archway

图 37 影壁起到的障景的作用
Fig 37 Screening wall

（5）结合基地的自然环境，进行总体的景观规划和设计，为总体规划锦上添花。

设计特点

（1）布局简洁明快，徽派建筑加中式园林，特点鲜明突出。（图24-26）。

（2）因地制宜（图27-30）。

（3）小中见大，充分发挥绿地的作用。

（4）植物配置与环境结合（图31-32）。

（5）生态性原则。一方面要达到植物生长与环境和谐统一的要求，以及植物群落的丰富性等特点，另一方面要提供特殊的阻隔、除尘、遮荫等防护性功能，并与水面、置石、道路、建筑等空间元素在时空间进行良好协调，达到植物生态习性、景观审美要求和整体空间意境的完美结合。

（6）动静分区为满足不同人群活动的要求（图33-37）。

总平面布置

无庶居住小区根据场地特点分东北区和西南区。

西南区：该区主要有建筑1、2、3幢，分别取名为降喜阁、见喜阁、登喜阁，成倒"品"字形，中央设置水景区域，形成中心景观。见喜阁西侧设置"还我读书处"，环境幽静，是小区内学子读书的好去处。登喜阁北侧设桐音馆、雅意亭，供小区内人们休闲娱乐，同时与小区东北区相衔接。

东北区：小区主入口位于西南区的右侧，向里延伸进入东北区，该入口做法别致，在入口处有一段景墙，开一门洞。进入小区要经过一个大堂，大堂中间摆放着一具假山石，人从两侧进入，而假山石正好把小区入口的景观遮挡住了，人只有通过大堂才能看到小区景观。东北区建筑主要由"禄"、"福"、"寿"三个建筑群组成，总体布局呈现较规则式，四周由绿地和小径环绕，楼与楼之间大尺度庭院楼台。"禄"南侧有桐音馆、雅意亭；"福"北侧设汲古得绠、濯缨亭、别有洞天等景点；"寿"南侧设濠濮间想、月到风来、观瀑，西侧廊、亭、轩与"福"相衔接。

道路交通系统设计

无庶居住小区的道路系统采用南北向干道从西南区的右侧穿入东北区，小区入口设于小区用地的南侧，由于现状条件限制，没有其他次入口，这也彰显了无庶的设计思想来源——"隐世"之梦。所谓"大隐于市"，无庶可为。西南区左端南侧有个地下车库，宽度7.5m，双车道行车。主干道的布置自然地形成了小区的分区。同时小区内没有机动车穿行，实现了"人车分流"，步行于园内人与自然最大化接触，真正达到物我两忘、养身养心的境界。

小区的步行道设计自成系统，相连为一体。一级步道宽3.6m为全园的主要步行干道；二级步道宽0.6-1.5m，作为居住区内内部与一级步行干路的联系道路。通过步行系统将区内的若干个景观节点和中心绿地串联起来，形成中心景区及景观带。

(4) To choose plants harmonious to the surroundings (Fig31-32)

(5) To be ecological: Plants should be in harmony with their surroundings and also clean the air of pollution, provide shade, and coordinate with the water, rocks, streets, and buildings in terms of space and time, becoming perfectly integrated with the whole artistic production.

(6) To vary between passive and active zones so as to meet the different needs of residents (Fig33-37).

The layout

The Wushu Residential District is divided into northeast and southwest areas.

Southwest area: Building 1, 2 and 3 are located here, their names are "*Jiangxi* pavilion", "*Jianxi* pavilion" and "*Dengxi* pavilion", they are arranged into the shape of the Chinese character "*pin*". A large pond is located in the center of the site to form the central landscape feature. To the west of the *jianxi pavilion*, is a spot called "A Place for Reading". It is a quiet and secluded place for children in the community to read books. To the north of the Dengxi Pavilion, is "*Tongyin House*" and "*Elegant Meaning Pavilion*". These structures are arranged for leisure and entertainment and also connect the north and south areas to the residences.

Northeast area: to the right of the southwest area, the main entrance extends into the northeast area. This unique entrance contains a scenic wall with a door.

The residential area is accessed by way of an alley where one passes through a rock feature. The rockery screens the view of the residential area. Only by passing through the hall can one view the interior scenery. The buildings in the north area are arranged in three groups called "*lu*" "*Fu*" and "*Shou*". The layout of these buildings is regular, with surrounding green spaces, small roads, and large-scale yards between the buildings.

Tongyin house and *Elegant Meaning Pavilion* are located to the south of the *Lu* buildings. A series of evocatively named features are located to the north of the *Shou* buildings, including: "*Acquisition from Ancestors*", "*Diying Pavilion*" and "*Journey into Amazing Caves*". Connected with the *Fu* buildings are: "*Thinking in Rivers*", "*A Visit from the Moon and wind*" and "*Waterfall Watching*".

Transportation system

The main north-south road extends from the entrance on the south, which emulates the ideas of "*The Dream of the Recluse*". This road creates the main divisions of the community. The underground parking area is located in the north area of the northwest zone.

No vehicles are allowed to cross through the community, thereby creating a separation between people and vehicles. This allows for a contemplative pedestrian experience where one can get in touch with nature, forget the outside world, and cultivating one's physical and mental states.

The pedestrian circulation system is integral. The 3.6m main road forms the central pedestrian throughway while the 0.6-1.5m secondary roads connect to the interior areas of the community.

作者简介：

江苏大千设计院有限公司 / 园林设计 / 江苏南京

Biography:

Jiangsu Daqian Design Institute Co.,Ltd. / Garden Landscaping / Nanjing, Jiangsu Province

查尔斯·沙（校订）

English reviewed by Charles Sands

贫民区的绿色生机
温卡特中心花园的设计与建设

A LITTLE BIT OF COUNTRY ON SKID ROW
DESIGNING & BUILDING THE WEINGART CENTER GARDEN

伊娃·卡纳普尔

Eva Knoppel

项目位置：美国加利福尼亚洛杉矶市
项目面积：5000 平方英尺
委托单位：温卡特中心协会
设计单位：伊娃花园景观设计事务所
景观设计：伊娃·卡纳普尔
完成时间：2010 年 6 月

Location: Los Angeles, California USA
Area: 5,000 square foot
Client: The Weingart Center Association
Designer: Garden of Eva Landscape Design Group, Inc.
Landscape Design: Eva Knoppel
Completion: June, 2010

图 01 改造前的温卡特中心花园
Fig01 Weingart Center Garden – Before

图 改造前的温卡特中心花园
Weingart Center Garden – Before

图 03 圣贝纳迪诺国家森林公园
Fig03 San Bernadino National Forest

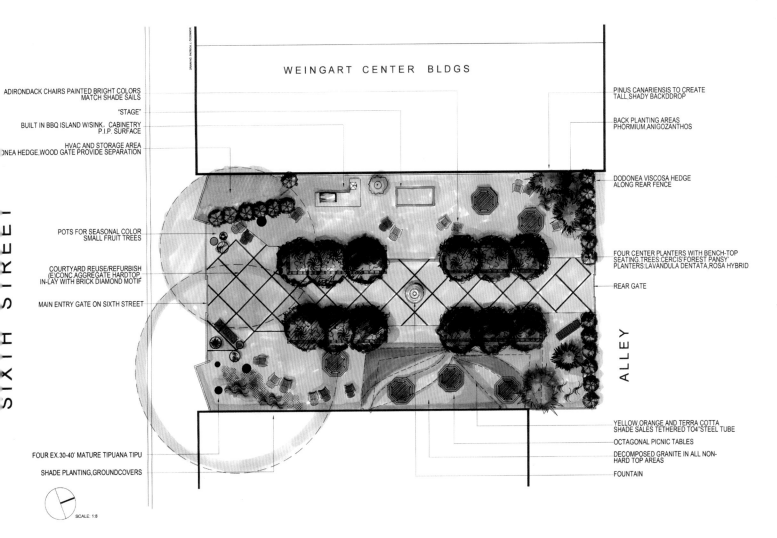

图 04 温卡特中心花园平面图
Fig 04 Weingart Center Garden Ground Plan

事情源于2010年5月某个周三下午的一通电话。电话那边问我是否愿意参与竞标为他们位于洛杉矶市中心总部旁边一块废弃的空地设计一座花园，尽管我从没听说过温卡特中心。他们还说要在下周一的中午与我会面，然后让我在2点之前就把设计和投标方案交给他们。我告诉他们会面是可以的，但是2个小时之内设计出一座花园然后把投标文件交到他们手里是不可能的。虽然我以工作效率高效著称，但这是要设计一座5000平方英尺的花园……所以，他们把日期宽限到周三。

当周一我们会面的时候，说实话我都不愿意下车。温卡特中心在洛杉矶位于一个众所周知的贫民窟——这附近以流浪汉、醉汉和瘾君子聚集而声名狼藉，而不是一个设计的花园好地方。我那时的想法是，"怎么会有人想让人在这种地方建个花园，而且为什么是让我来设计？"但是我还是下车了，我现在特别庆幸我下车了！

温卡特中心花园

我从他们的网站 (www.weingart.org) 了解到，"温卡特中心协会是一个提供多种服务的机构。它负责提供高质量的人性化服务，在帮助打破无家可归的恶性循环和消除贫穷方面起领导作用，并提供创新性的解决方法。它的项目为退役老兵、妇女、假释人员、艾滋病病毒携带者、流浪汉、以及濒临无家可归者服务。他们的目标是帮助受助者脱离无家可归和自我毁灭的境地，协会通过教会他们技能，发放必要的工具和资源，以及传递希望来让受助者过上自己自足的生活。"中心现在设在 El Rey 酒店的原址，一座1926年建成的十一层大楼。建造花园的选址就位于既能满足中心使用又能为他们服务客户的地方。

场地

我不知道这里之前是做什么的，现在它被一条通往中心的水泥砖人行道上的沥青所覆盖（图01）。沥青必须清理走，但是人行道可以利用，毗邻楼房的红砖墙也可以被利用（图02）。

我作为景观设计师从事设计行业已有18年了，期间完成的1000多项工程足以使项目的选址和它们周边的建筑物给我灵感，引导我的设计构思，就好像它们可以对我对话一样。当我在这平民窟废墟之间走动的时候，我意识到只把它改造成一个精致漂亮的袖珍花园。我不断地问自己："如果我能成为什么的话，那我将会成为什么？"然后我就想到了我和孩子们在大熊湖的时候，那是一片由积雪融化形成的美丽湖泊，被圣贝纳迪诺国家森林公园所环绕，（图03）这里位于洛杉矶东北部大约100英里的地方。孩子们喜欢在森林中奔跑穿梭，于是我想到我是不是也可以让建造的这个地方给贫困环境中挣扎的人们带去些什么。那将是给他们提供一个能与大自然亲近的地方。于是，这个花园诞生了。

过程

我既是一名景观设计师，也是持证承包商，通常自己建造自己设

It was a Thursday afternoon in May of 2010 when the call came. Although I had never heard of the Weingart Center, they called to ask if I would bid on designing a garden in downtown Los Angeles on a derelict piece of property next to their headquarters. They wanted to meet with me the following Monday at noon, and wanted my design and bid by two o'clock that afternoon. I told them I could make the meeting, but there was no way I could design a garden and have a bid in their hands in a matter of two hours. I have a reputation for working quickly, but to design a 5,000 square foot garden . . . so, they gave me till Wednesday.

When I arrived for Monday's meeting I didn't want to get out of my car. The Weingart Center is in an area of Los Angeles known as skid row—a run-down neighborhood more famous for vagrants, alcoholics and drug addicts than for landscaped gardens. All I could think of at that moment was, "Why on earth would anyone want to build a garden here . . . and why me?" But I got out of my car, and I'm glad I did!

The Weingart Center

As I learned from their website (www.weingart.org) "The Weingart Center Association is a multi-service agency that delivers high-quality human services and provides leadership and innovative solutions to help break the cycle of homelessness and end poverty. Its programs are tailored for veterans, women, individuals on parole or that are HIV positive, and for the homeless or those at risk of becoming homeless. The Center's goal is to provide men and women with the skills, the tools, the resources and the hope they need to help break the cycle of homelessness and self-destructive behavior and lead self-sufficient lives."

The Center and its clients reside in what was originally the El Rey Hotel, an 11-story building constructed in 1926. The piece of property they wanted transformed into a garden would serve both the Center and its clients.

The Property

I had no idea what the property was originally used for, but it was now covered in asphalt with a concrete and brick walkway running down the center (Fig.01). The asphalt had to go, but the walkway could be used, as could the red brick wall of the adjacent building (Fig 02).

I've learned over the course of 18 years as a landscape designer and completing well over 1,000 projects to allow the

图05 美国志愿队志愿者们参与建设温卡特中心花园
Fig05 Weingart Center Garden Construction with AmeriCorps Volunteers

图06 温卡特中心花园建设之中——花池的准备工作
Fig06 Weingart Center Garden Construction – Planter Preparation

图 07 遮阳膜设计
Fig 07 Shade Sails Design

计的项目。然而，这个项目却不是这样。温卡特中心协会加入了美国志愿队 (www.nationalservice.gov/programs/americorps)。美国志愿队是一个非盈利组织，每年为成千上万的年轻人创造就业机会，以使他们获得有益的工作技能、赚取学费，提高作为一个市民的荣誉感。将有一组志愿者参与到花园原址的清理和花园的建设中来。

设计

需要改造的场地是一块94英尺长，55英尺宽的土地。温卡特中心在它的一侧，一座用砖砌成的建筑物在另一侧。中间的水泥砖路从第六街开始到另一头的一条小路和停车场结束，两侧都被10英尺的栅栏和中央大门围合起来。

按照温卡特中心的要求，建造的花园既可以用作休闲聚集场所，也要可以承办大型活动。它需要容纳200人，还有桌椅，烧烤设备，冰箱，洗脸水池，所有的设备都要易于维护。

我决定保留中间的人行道，把整个地方分成4个部分。（图04）左边的这一个部分，也就是离第六大道最近的这一部分，有固定的供暖和空调设施，将用混凝土浇筑起来。"L"型的岛型区域，将放置烧烤设备，冰箱和洗手池，还有一大片地方用作餐饮服务的场所。横穿人行道的这片区域被设计成一个休闲休憩场所，有椅子、长凳和野餐桌等设施。这块场地的另一头，靠左的部分将建起一个升高的平台，可以被用作舞台或舞池，靠右的那部分则建设成摆放野餐桌椅的地方，上面则有绿藤覆盖的廊架以提供阴凉。

为了建造能容纳需要数量座椅的场地，我设计了四个在矩形花池的四周长椅（22'x7'），它们由水泥砌成，上边铺设一层Ipewood（一种巴西胡桃木）作为坐的地方。矩形长椅围合出来的花池里栽种着玫瑰、薰衣草和紫荆。游客进入花园，林荫道就会把游客的视线指引到公园的另一头。一座三层结构的喷泉位于四个矩形长椅的中央，也是花园四个部分的中心，即整个花园的中心。

要使花园给人一种"田园"的感觉，人行道之外的地面用风化的花岗岩铺就；为了掩盖花园后面的停车场，设了一道树篱屏障，还在人行道的两头分别种植了两棵松树；公园里的设施包括了手工制作的阿第伦达克椅（一种户外沙发，靠背可以调整角度，沙发座通常用宽的长木条制成），野餐用桌和木质长凳，所有的这些设施都将被粉刷成黄色，红色或者橙色。

建造

所有事情都进展顺利：沥青已经被清理干净，红砖墙也被清洗一番，排水系统的建造也完成了，混凝土都浇筑完成了，只是建造廊架的许可证至少6个月之后才能获得，因为建造花园的这块地皮从来没有被纳入到城市审查中来，建造的许可证只能发给经过审查的土地（图05-06）。

我几乎不可能等待六个月后才拿到许可，所以我决定放弃建造廊架而利用六根独立的支柱。我找来暗黄色和橙色的帆布，把它们裁成三角形。帆布在支柱之间通过长绳搭建起来，形成了一大片阴凉地，这也给整个公园带来了一种节日的气氛（图07）。

property and the associated buildings to guide me in the design process. It's almost as if they speak to me. What I heard as I walked about that forlorn slice of skid row was that it wanted to be anything but a pretty little pocket park.

I kept asking myself ... "If I could be anything, what would I be?" Then, I remembered the time my kids and I were up at Big Bear—a beautiful snow-fed lake surrounded by the San Bernardino National Forest, (Fig.03) about 100 miles northeast of Los Angeles. They loved running through the forest and I decided that if I could bring anything to the people who were struggling here in this hardscrabble environment, it would be to provide them with a little bit of the out-of-doors ... and so a garden was born.

The Process

As I am both a landscape designer and licensed contractor, I usually build what I design. However, that was not going to happen with this project. The Weingart Center Association had joined forces with AmeriCorps (www.nationalservice.gov/programs/americorps) a not-for-profit service organization that each year creates jobs for thousands of young adults in order to help them learn valuable work skills, earn money for education and develop an appreciation for citizenship. I would have a crew of AmeriCorps volunteers to clear the property and construct the garden.

The Design

The property to be transformed was 94-feet long and 55-feet wide. The Center was on one side and a brick building on the

图08 温卡特中心花园
Fig08 Weingart Center Garden

图09 温卡特中心花园
Fig09 Weingart Center Garden

图 10-11 温卡特中心花园
Fig10-11 Weingart Center Garden

图 12-13 温卡特中心花园
Fig12-13 Weingart Center Garden

一切都按照计划进行：滴灌和照明系统安装就绪，周边的区域也都栽种上了地中海植物和耐旱植物，比如番石榴，迷迭香，亚麻和海桐（图08-13）。

美国志愿队的志愿者们做出了非常了不起的工作，他们为这个花园和自己的劳动成果而自豪。

结语

公园的开放是个巨大的成功。洛杉矶市的市长安东尼奥·维拉让哥萨出席并发言，对花园的建成付出心血的来自不同地方合作机构的嘉宾也到场出席。但对于我来说，值得纪念的却是大家给与的回应——温卡特中心居民和附近邻居对新花园表现出的快乐，骄傲和兴奋。

我设计过很多数价值数百万美元的景观，但是没有一个能够比拟设计这个使得贫民窟焕发生机的花园所给我带来的满足感。□

---

other. The bisecting concrete and brick pathway ran from Sixth Street at one end to an alley and a parking lot on the opposite end, and was enclosed at both ends by 10-foot fencing with central gates.

The Center's list of requirements spelled out that the garden was to be used for both casual gatherings and large, organized events. It needed to have seating for 200 people, as well as tables and casual chairs, a barbeque, refrigerator, sink with running water, and be easy to maintain.

Having decided to keep the central walkway, I divided the property into four quadrants. [Fig.04] The left quadrant, closest to Sixth Street, which had a heating and air conditioning unit that could not be moved, would have a poured concrete, "L-shaped" island. The island would house the barbeque, refrigerator and sink, and its top would provide a large surface for food service. The quadrant across the walkway was designated for casual seating and would contain chairs, benches and picnic tables. At the other end of the property, the quadrant to the left would have a small raised platform that could be used as a stage or a dance floor and the quadrant to the right would have picnic tables and chairs with a large pergola covering it for shade.

To accommodate the required amount of seating, I designed four rectangular planters (22' x 7') to be built of concrete with Ipewood (Brazilian walnut) decking for the sitting area. The planters flanked the concrete and brick path and were filled with roses, lavender and Cercis Forest Pansy (Western redbud) trees. When one entered the park, the trees served to create an allée that led the eye to the far end of the park. A three-tiered fountain was at the center point of the planters, the quadrants and the garden.

As the feeling of the park was to be "country," decomposed granite was used for the surface of areas not covered by the walkway. Across the back of the park, to mask the parking lot, a hedge was installed and two pine trees planted on either side of the walkway. The furniture would include hand-made pine Adirondack chairs, picnic tables and wooden benches and all would be painted yellow, red or orange.

The Construction

All was going swimmingly: the asphalt had been removed, the red brick wall had been power washed, all the drainage had been completed and the concrete was being poured when I was notified that processing the requested building permit needed to construct the pergola would take at least six months because this piece of land had never been included in the city's survey and a building permit could only be issued on surveyed property (Fig.05-06).

I could hardly wait six months to get the permit, so out went the pergola and in went six freestanding stanchions. I had sailcloth died yellow and orange and cut into triangles. The sailcloth was run on guy wires between the stanchions, which created overlapping areas of shade, giving the entire park a festive feeling(Fig.07).

Everything else went as planned: the drip irrigation and lighting were installed and the perimeter area planted with Mediterranean and drought-tolerant plants such as guava, rosemary, flax and pittosporum (Fig.08-13).

The AmeriCorps volunteers did a tremendous job and were extremely proud of the garden and what they had accomplished.

The Finale

The park's opening was a great success. Los Angeles Mayor Antonio Villaraigosa spoke and everyone from all the agencies and organizations who made the garden possible were there. But the take-away for me was the response—the joy, the pride and the excitement—the Weingart Center residents and their neighbors had for their new garden.

I've designed any number of stunning landscapes for multi-million-dollar estates, but none of them gave me half the satisfaction that I got from bringing a little bit of country to skid row. ■

作者简介：

伊娃·卡纳普尔／女／住宅和商用景观设计与建筑商／伊娃花园景观设计事务所／美国加利福尼亚洛杉矶

Biography:

Eva Knoppel / Female / Residential & Commercial Landscape Designer & Contractor / Garden of Eva Landscaping Design Group / www.garden-of-eva.com / Los Angeles, CA USA

昆山夏驾河"水之韵"城市文化休闲公园（廊架小品）
——2013全国优秀工程勘察设计行业园林景观三等奖，2013年度上海市优秀工程设计一等奖

上海亦境建筑景观有限公司以规划、建筑、景观专业的设计实力为依托，构建了从项目前期策划、整体规划、建筑设计、景观设计到景观工程的一体化服务体系，为客户提供项目前期至后期的全方位解决方案。

亦境公司汇聚百余名设计精英、驰骋行业二十年，拥有建筑行业乙级、风景园林专项甲级、城市园林绿化二级等行业资质，作品多次获得国家建设主管部门及上海市优秀工程设计奖项。

大型城市滨水区
昆山夏驾河"水之韵"城市文化休闲公园规划设计
昆山夏驾河滨水商业娱乐区规划、建筑、景观一体化设计
泰州凤城河-西南城河景观、建筑一体化设计
镇江古运河风光带总体规划及分段设计
镇江国宾馆建筑、景观、夜景灯光一体化设计（金山湖）
镇江滨江风光带规划、建筑、景观、灯光一体化设计
海南博鳌香槟郡规划与建筑设计（滨海综合片区）
镇江内江滨水城市景观规划与设计

农业科技园区
上海农科院奉浦院区规划、建筑、景观一体化设计（合作设计、景观施工）
天津市现代农业科技创新中心（西青）规划及建筑设计
天津市现代农业科技创新基地（武清）规划及建筑设计
鄱阳湖生态经济区现代农业科技创新示范基地规划及建筑设计
南昌市溪霞现代农业科技示范园规划、建筑、景观一体化设计
安徽省农业科学院（本部基地）规划设计

古典园林与建筑
上海东海观音寺建筑及环境设计
上海醉白池公园景观改造设计
扬中市佛教文化广场规划设计

道路绿化景观
泰州市长江大道景观设计
泰州溱湖大道景观改造设计
张家港金港大道景观提升设计
江阴霞客大道景观设计

居住社区
上海青浦重固逸皓华庭景观规划设计工程（设计、施工一体化）
春申景城景观绿化工程（景观施工）
溧阳天目湖健康养生园景观规划设计
靖江新桥别墅区规划、建筑、景观一体化设计
博鳌亚洲论坛国际社区规划、建筑设计

EDGING 亦境建筑景观

ADD：上海市普陀区中江路388号国盛中心1号楼3001-3003 （PC:200062）
Rm. 3001-3003 Gouson Center Building 1#, 388 ZhongJiang Rd, Putuo District ,Shanghai
TEL：021-6167 7866（总机） FAX：021-6076 2388
http://www.edging.sh.cn

建筑设计研究院 E-mail：arch@edging.sh.cn QQ：12 9321 8362
景观规划设计院 E-mail：la@edging.sh.cn QQ：19 5293 4808
景观工程公司　　E-mail：gc@edging.sh.cn

注：数据均由上海亦境建筑景观有限公司统计

亦小亦美·亦真亦善

济南市园林规划设计研究院
JINAN LANDSCAPE PLANNING AND DESIGNING RESEARCH INSTITUTE

济南市园林规划设计研究院于1983年11月22日正式成立。拥有园林规划设计甲级、建筑设计乙级资质。院下设总工室、五个园林规划设计所、一个建筑设计所、效果图表现室、ST环境设计研究所（与日本景观设计大师德永哲先生合作成立）、济南彩叶园林工程项目管理有限公司等部门。我院所做设计项目获得部省市各级奖励100余项，其中国际奖项及国家级奖项16个，部级奖项7项，省级奖项24项。我院强调"虽由人做、宛自天开"源于自然而高于自然，把握传统园林的精髓，开阔思路，形成独具一格的设计风格。

传　真：0531-82970090
电　话：0531-82053043(办公室)
　　　　0531-82971782(经营部)
地　址：山东省济南市市中区六里山路10号
网　址：http://www.jnlad.com

Oasis 无锡绿洲
景观规划设计院
有限公司

无锡绿洲景观规划设计院有限公司成立于2004年，具备建设部颁发的风景园林设计专项甲级资质。我们的宗旨是在不同的专业领域中，力求景观设计的功能性、创新性、人性性以及环保性。绿洲坚持运用当代设计手法及语言，将自然、人性与艺术作为不懈探索的设计命题，以务实的态度和高度的热情参与实践。

我们始终注重博采众长，不断创新，并且通过我们与客户之间的合作，建造可持续发展的环境。在城市公园、绿地及水系、风景旅游区、住宅及商业区等领域的规划设计中，提供了独特的解决方案和优秀的服务品质，得到了广泛的认可。

Landscape Design	Urban Planning	Architecture	Environment
景观设计	城市规划	建筑设计	环境咨询

WUXI LVZHOU

Urban Planning
城市空间景观设计

Commercial/Business District Design
商业/办公区景观设计

Hotel Design
酒店景观设计

Residential Design
住宅区景观设计

School Design
学校景观设计

Attractions Design
景点景观设计

地址：江苏省无锡市滨湖区湖滨街 15 号蠡湖科技大厦 23 楼
电话：0510-66968096　0510-66968087
传真：0510-66968091　邮箱：lvzhou215@163.com
网址：www.wxlvzhou.com

对话大师
MASTER'S DIALOGUE

亨利·巴瓦   **Henri Bava**

亨利·巴瓦教授是法国著名风景园林师，也是原法国景观师协会会长。先后担任法国凡尔赛国立高等景观学院、德国卡尔斯鲁厄理工学院教授、哈佛设计学院景观系客座教授，并多次受邀到宾夕法尼亚、巴塞罗那、哥本哈根、墨尔本等多所大学进行讲学。他所经营的事务所曾多次获得欧洲的多项景观及规划奖项。

Henri Bava Professor Bava is a renowned landscape architect in France and he has served as the former chairman of Association Des Paysagistes Conseils de l'Etat. He has previously taught at Ecole nationale superieure du paysage de Versailles, Karlsruhe Institute of Technology and Department of Landscape Architecture Design College, Harvard University as a visiting professor. In addition, Professor Bava has been invited on numerous occasions to lecture at different universities (such as University of Pennsylvania, Barcelona, Copenhagen, Melbourne and etc.) around the world. Not only that, the architectural firm that Professor Bava operated has received numerous European awards for landscape projects and planning.

对话大师 MASTER DIALOGUE

图 01 亨利巴瓦照片
Fig 01 Henri Bava couleur

图 02 亨利巴瓦及其合伙人
Fig02 Henri Bava and his partners

对话大师 MASTER DIALOGUE

03 工作室
g03 The photography of the studio

2013年1月12日晚，第三届园冶高峰论坛之风景园林师沙龙于北京新大都饭店举行，沙龙邀请了欧洲著名景观设计师亨利·巴瓦先生与来自国内的众多知名景观设计师，进行面对面的交流和探讨。

此次风景园林师沙龙主持人为世界园林杂志总编王小璘教授，出席沙龙的嘉宾有西安市市容园林局副局长吴雪萍女士，清华大学建筑学院景观学系副主任朱育帆教授，华中农业大学副校长高翅教授以及易兰国际设计公司副总裁唐艳红女士。

王小璘（以下简称「王」）：各位嘉宾，大家晚上好。非常高兴能跟大家一起进行大师对谈。今天晚上的大师是在今天下午一场非常精彩报告的亨利巴瓦教授，在座的几位嘉宾是园林界的龙头，通过嘉宾和各位与大师的对谈，相信今天会有一场非常精彩的交流会。

首先介绍亨利巴瓦教授。巴瓦教授是法国著名风景园林师，也是原法国景观师协会会长，先后担任法国凡尔赛国立高等景观学院、德国卡尔斯鲁厄理工学院教授、哈佛设计学院景观系客座教授，并多次受邀到宾夕法尼亚、巴塞罗那、哥本哈根、墨尔本等多所大学进行讲学，他所经营的事务所曾多次获得欧洲的多项景观及规划奖项。接下来介绍几位与会嘉宾：

吴雪萍局长，西安市市容园林局副局长，总工程师兼西安市古建园林设计研究院院长；

高翅校长，华中农业大学副校长；

朱育帆教授，清华大学建筑学院景观学系副主任；

唐艳红女士，易兰国际设计公司副总裁。

欢迎几位嘉宾的到来，我是今天晚上的主持人王小璘，很荣幸有这个机会跟大家一起来进行今天晚上的大师对谈。

这次大会的主题是「魅力中国」，主讲人是亨利巴瓦教授，他的报告为我们做了一个非常系统的整理，将景观领域分为5种类型。而通过他这几年来在国际上非常多的实际案例，我们也看到他涉猎的触角非常广，包括住宅、工业、立体绿化等等。他不但在拥挤的城市有限的空间里创造了很多的绿化空间，并且对从地面到地下的分层景观做了非常严谨而有深度的分析。同时，他也让我们理解到，景观不仅是传统的视觉美学而已，它还包含更深层的内涵：一个是文化，一个是生态，还有一些经济带动的作用。亨利教授的演讲传达给我们的一个讯息，景观可以带动一个城市，城市需要好的景观来带动，城市也要去引导景观。这是我从亨利巴瓦教授的精彩报告里得到的一些心得。

接着，请教亨利巴瓦教授，很多时候由于人为的介入和破坏，对环境造成了不良的影响，比如环境污染。我看您的很多项目里都考虑到了生态问题，但是在不同的国家或者是不同的领域，甚至于不同的人对生态的解读是不太一样的，有的会从文化生态的角度去看它，有的会从自然生态的角度看它，包括野生动植物的栖息地等。请问亨利巴瓦教授，您对生态的解读，或说是定义，比较偏重在哪方面？为什么这部分会比较重要？

亨利巴瓦（以下简称「巴瓦」）：所有场地的基本元素，包括地质、水文等，都是非常重要的，不能说哪一种是最重要的生态元素。我们考虑场地的时候，不能单独的考虑某一种元素，而是要综合的来考虑场地的整体，包括场地边界以外的各种元素是如何影响我们的园林规划和景观设计的。

解读两个比较重要的概念，第一是景观的概念，第二是环境的概念。就欧洲景观的发展历程来讲，环境跟景观是完全不一样的，如果我们用带有一种欣赏的眼光去看环境的时候，这个环境就不再是环境，而是一种景观。所以就看我们当代的景观设计师如何去考虑，如何依据每个城市本身具有的特点来考虑当地特色的景观。

王：换句话说，城市的特色应多考虑文化层面，在设计之前要先去了解城市的特质，这能不能解释成是一个城市本身的文化？

The 3rd Landscape Architect Salon was held at Capital Xindadu Hotel Beijing in the evening of January 12th, 2013. Henri Bava, a renowned landscape designer from Europe and several prominent domestic landscape designers were invited to the event for a session of face-to-face exchange and discussion.

The salon was hosted by Professor Xiaolin Wang, the Honorary Chairman of the Chinese Landscape Architects Society in Taiwan and editor-in-chief of the Journal of Worldscape. Other guests invited to attend the event include Xian City Landscape Bureau, Deputy Director Ms. Xueping Wu, Professor Yufan Zhu, Associate Director, Institute of landscape Architecture, School of Architecture, Tsinghua University, Professor Chi Gao, Vice-President of Huazhong Agricultural University and Ms. Yanhong Tang, Vice-President of ECOLAND Planning & Design Co., Ltd..

Xiaolin Wang (Name as 'Wang' below): Good evening, everyone. It is my great pleasure to be able to participate in the Masters' Dialogue with you here tonight. The Master we have here is Professor Henri Bava, who just gave an impressive presentation this afternoon. Our guests here are also heavyweight figures in the landscape architecture field. With the dialogue between the master, guests and the audience, I am confident that we will have a very informative and fruitful exchange in this salon.

First, allow me to briefly introduce Professor Henri Bava. Professor Bava is a renowned landscape architect in France and he has served as the former chairman of Association Des Paysagistes Conseils de l'Etat. He has previously taught at Ecole nationale superieure du paysage de Versailles, Karlsruhe Institute of Technology and Department of Landscape Architecture Design College, Harvard University as a visiting professor. In addition, Professor Bava has been invited on numerous occasions to lecture at different universities (such as University of Pennsylvania, Barcelona, Copenhagen, Melbourne and etc.) around the world. Not only that, the architectural firm that Professor Bava operated has received numerous European awards for landscape projects and planning. Now let me introduce the other participating guests:

Deputy Director Ms. Xueping Wu of Xian City Landscape Bureau. She is also the chief engineer and director of Xian City Ancient Architecture & Landscape Design Institute;

Professor Chi Gao, Vice-President of Huazhong Agricultural University;

Professor Yufan Zhu, Associate Director, Institute of landscape Architecture, School of Architecture, Tsinghua University,

And Ms. Yanhong Tang, Vice-President of ECOLAND Planning & Design Co., Ltd..

I would like to start off the evening salon by welcoming our guests here. My name is Xiaolin Wang and I am the host for the event, and it is my honor to be here with everyone tonight for the Master's Dialogue.

The theme for this the forum today is "Charming China", with Professor Henri Bava as keynote speaker. His speech offered a very systematic organization of landscape domains into five types. Through the actual international projects that he had been involved in the past few years, we can see that Professor Bava has delved into a number of different fields, including residential, industrial, 3D greening and so forth. Professor Bava has not only created many green spaces in limited areas of densely populated

图 04 林中居 1
Fig04 Bois Habite1

图05 林中居2
Fig05 Bois Habite2

巴瓦： 刚才王教授说景观是一种文化，如果想了解当地的景观文化，必须要真正的生活在这个场地内。比如说在法国的一个岛屿，我们作了一个项目，首先我们会去了解这个场地具有的特质，我们把工作室建在那边，真正生活在那个城市里，了解当地人的文化和当地生态的景观特质，来作为项目的出发点和它的灵感来源。如果不这样做的话，这就只是一个表面的形式作业。如果我们想真正理解场地，必须与它相爱，真正去了解它的内在，了解它的本质。

王： 因此，景观或城市就必须具备文化的特质，才能呈现具有深度的意涵。接下来请哪位嘉宾来提问？

高翅（以下简称「高」）： 我从亨利巴瓦的作品里读到设计哲学，在今天的报告里，他把景观分成五类，所有作品对于场所的文脉和地脉，也可以说是自然系统和文化系统，予以充分理解基础上的尊重。实际上这是在尊重他人、尊重历史的同时也尊重了自己，这是我读出来的哲学。

我的第一个问题，是今天下午看到很多照片，在实景照片里都没什么人，在效果图里面却有很多人，我的问题是，那个照片是一定要选在没有人的时间，还是拍的时候没有人？

巴瓦： 这些照片不是公司去拍，而是邀请专业摄影师去拍。他不是专业的人员，他不知道场地的活力在什么地方，他认为这个地方没

cities but has also made a meticulous and in-depth analysis of stratified landscapes from underground level to floor level. He has also helped us understand that landscape is more than just traditional visual aesthetics and that it embodies significance at deeper levels, namely culture, ecology and facilitation of economy. Professor Bava's speech has delivered the message that while landscape could drive a city forward and that cities require quality landscape as propellant, cities must also serve as a guide for landscape architecture. This is what I have learnt from Professor Bava's wonderful report.

And now I have a question for Professor Bava: We have seen from numerous instances that the intervention and destruction caused by man has brought adverse impact (such as environmental pollution) upon the natural environment. I noticed in many of your projects have taken ecological issues into consideration. However, different countries, different domains and even different demographics could have varying interpretations on ecology. Some see it from the perspective of cultural ecology while others approach the topic from the standpoint of natural ecology (including the habitat for wildlifes). With that said, what would your interpretation (or definition if you will) of ecology and what aspect of it would you emphasize? And why?

6 林中居3
6 Bois Habite3

有人,这个地方才是最漂亮的,所以他就拍这样的照片。这是公司与摄影师之间的一个错位,是他们在交流上的问题。

我们也不能只是关注在这个场地里游人的存在,其实这个场地是由很多人的共同作用来形成的,这种变化也是由居民、游客的共同作用之下,在不断的演变之中形成的。我们不能只看照片中有人,就觉得这个地方富有活力,或者没有人就说它没有活力,其实很多共同因素相互交叉,造成这个场地不断的演变,不光是由场地中的游客们和散步的人们来决定的。

欧洲的景观设计不仅是由景观师或者建筑师来决定,这块场地将来会是什么样子,是由存在在这个场地里的人和居民共同来提出这个场地的优缺点,他们和景观师一起讨论,提出对于未来生活环境的想法。以这样的结果来决定他们将来的设计是什么样的,而不是单纯的按照景观设计师的想法执行。

王:我们称之为公众参与。其实公众才是开放空间的使用者,而不是由景观师来决定场地以后应该成为一个什么样的景观。所以亨利教授给我们的信息是景观师在做设计的时候,一定要尊重使用者或者是潜在的使用者,比如说在设计住宅景观的时候,住宅景观的屋顶做了绿化,这个部分肯定要有居民的参与;否则,屋顶花园横跨很多住宅,有的会喜欢,有的不一定会喜欢,景观设计不是我要给你什么?而是要知道,你想要什么?

Henri Bava (Name as 'Bava' below): The fundamental elements of any site, such as geology and hydrology, are all very important. It would be inappropriate to identify a specific element to be the most important ecological factor. In the consideration of sites, we cannot simply focus on just a single element alone; we have to make a comprehensive assessment of the site based on all elements involved, including other factors beyond the perimeters of the site that could affect the process of garden and landscape designs.

There are two important concepts that require proper interpretation, the first being the concept of landscape and the other is the concept of environment. If we consider the developmental history of European landscapes, it is apparent that "environment" and "landscape" are two entirely different things. When we perceive environment from an appreciative point of view, environment would be more than just natural environment; it would become a landscape. So it is up to contemporary landscape architects like us to create landscapes with distinctive territorial features by taking unique features of each city into consideration.

图07 林中居4
Fig07 Bois Habite 4

巴瓦：对的，法国、德国、西欧国家做景观的方式已在慢慢转变，他们非常重视生活在场地里的居民，尊重居民的意见，专业景观师和居民进行交流，共同来完成这个项目，而不是由景观师单独的、武断的决定未来的设计。

高：还有一个问题。下午说到塞纳河岛上一个项目，是一个矩形的公园。我想知道公园的形态、面积以及位置是根据什么来确定的？

巴瓦：这个公园在在巴黎西部，叫做比浪谷，是一个雨水花园。花园的面积和形态不是由我们决定的。有时候做一个城市重大项目，我们可以决定我们所设想的形态包括面积，但这一次不是，因为这一次是一个综合项目，建筑设计师、城市规划师、景观设计师通力合作，进行整个区域的研究，建筑设计师和规划师对城市的区域规划做出了定义，给予花园大概的形态、位置和面积，我们接受到这个任务而去设计这个花园。在法国，景观设计是相当传统的，甲方给你一个任务，然后你来设计。当然他们的设计模式跟中国还不太一样，在中国，公园的面积和形式是甲方给他们定好的。

王：显然的，亨利巴瓦教授强调的就是设计者跟使用者之间应该要有所互动。

Wang: In other words, one should also consider cultural aspects of urban features and learn more about a city's characteristics prior to the design process. Can such characteristics be perceived as part of the city's culture?

Bava: Professor Wang raised an interesting point by referring to landscape as a type of culture. In order for one to truly understand the landscape culture of a place, he must actually live in the area. For example, we worked on a project on an island in France. We started off the project by learning the special characteristics of the place and we accomplished this by constructing our studio there so that we can actually live in that city. That allowed us to better understand the culture of the local population and the characteristics of local ecological landscapes so that we had a starting point for the project and a source of inspiration for ideas. If we had not done that, the project would be reduced to nothing more than a superficial formality. If we want to truly understand a site, we would have to fall in love with it in order to fully understand the site's inner beauty and essential qualities.

朱育帆（以下简称「朱」）: 今天中午我是在饭桌上问过巴瓦教授一个问题，作为一个欧洲人，你对美国怎么看？他说了一个字，大。美国相对欧洲来说，文化是有问题的，这两年因为教学的原因，我对欧洲的访问也比较多，对文化的认识也在深化。今年我去了一趟西班牙，跟西班牙有一个合作，我跟那里的教授探讨文化特质的问题，有两点非常有感触。

我去参观了一个建筑师的工作室，是后现代的一个建筑大师，他的工作室是1972年一个废旧水泥厂改造而成的，我觉得他做了二0年以后拉茨做的工作。第二个，我参观了欧洲景观奖的一个最佳设计，把原来弗兰克时期的遗址，变成了一个纪念遗址，这让我有一个非常深刻的感触，原来旧城改迁剩下来的遗址，也会变得这么漂亮。

巴瓦: 关于欧洲景观和美国景观，比较两者之间的优缺点还是比较困难的，我们知道美国景观有一个比较著名的案例，是詹姆斯做的一个项目。

朱: 欧洲景观对于场地特质的把握是有显性的价值，原来可能是由设计师关注一个明显的特质，现在变成由设计师去激活其中有潜质的东西。刚刚的案例也很优秀，如果我们真的做到完美，也就不需要跟你交流，我肯定发现了自身的问题，所以才要进行交流。我想跟亨利巴瓦教授请教的是，欧洲在这样价值观转化中的过程，从一个欧洲人的角度来谈一下是怎样进行转变的？欧洲设计师越来越重视文化潜质价值的开发，他们会很小心的进行保留，我想了解一下，从他的角度来说，我并不认为欧洲是完美的，但是我希望他能阐释出这样一个概念。

巴瓦: 从欧洲景观发展历程来看，我们所有人即便不是景观设计师，不是在这个专业范畴内的人，当我们看到一个很美的景观，我们都会赞叹。但是现在的情况是，欧洲景观设计师更注重的是我们看不到的东西。我们会认为这个景观有问题和缺陷，就会更对这些东西感兴趣。下午汇报的时候，我提到德国两个城市之间的矿厂公园，法国人觉得工业遗产不需要保留，因为法国把这种工业遗产都拆掉了。但是德国人就觉得这个工业遗产很漂亮，我们必须要保留下来，这是两国文化的不同点。法国人从德国人那里学习到有些东西可能需要保留下来。

在我法国的公司里，我们永远不能说："漂亮"和"不漂亮"，我们不能评价这个东西是漂亮还是不漂亮，场地本身就是以这些具有特色的东西为基础来进行研究和设计，而不是说直接判断它是美或者不美，在这个行业当中是严格禁止这样说的。

在1980年之前，法国和德国的设计，很多都是我们来找一种原形，比如说从法国园林或者英国园林，找一种原形来套用于这个场地。但是最后发现这个场地有一些需要去挖掘的东西，所以我们就不能单纯去套用这些东西，而是要去研究这些场地。

1980年之前，欧洲景观作品的价值高于场地本身，因此，我们做出来的作品要高于这个场地，但是在1980年之后，我们觉得场地本身的价值高于作品，所以要更深入的去挖掘他们本身内在的东西。一旦能够把作品跟场地两者之间完美的结合起来，这将是非常完美的，也是我们所追求的。

王: 刚刚朱院长提的这个问题非常好，欧洲各个国家有它民族性的特质，今天我们讲的欧洲是很大范围的，但每个国家却有自己的文化特色。回应朱院长所提的问题，亨利教授从德国那边学到了怎么去尊重别人没有看到的问题，试图把它做得更好，德国对废旧矿坑，有把它保留下来的意向存在。在过去没有特别强调这部分，这也是刚刚所说的某种文化特质。亨利教授在全世界都有项目，他看得多，知道自己需要学习的是什么？知道自己可以从别的国家学到什么。透过国际的交流，可以分享一些经验和心得。

Wang: This means landscapes or cities must have cultural characteristics in order to achieve in-depth presentation of meaning and significance. Who would like to be the next to raise a question?

Chi Gao(Name as 'Gao' below): I have read about design philosophy from Professor Bava's works. In his presentation today, he divided landscape into five specific types. All his works had demonstrated sufficient understanding and fundamental respect for all the sites' contexts and key lines (i.e., cultural and natural systems). In reality, such respect for others and respect for history reflect respect for oneself. That is the philosophy I have gathered from Professor Bava's works.

My first question comes from the pictures that I saw earlier this afternoon. In those pictures, I noticed very few people in them. But in the design sketches, there were many people illustrated in the images. So my question is, were those pictures taken intentionally during periods of few people or the total absence of people or was that just a coincidence?

Bava: To clarify, those pictures were not actually taken by the company; we invited a professional photographer to take those photographs. Unfortunately, he was not a professional landscape architect and he did not know the spot of vitality in those sites. To him, a place is most visually impressive when it is devoid of people, which is why he took pictures like those. It is an instance of misalignment between the company and the photographer – something went wrong in their communication.

On the other hand, it is important that we focus on just the presence of visitors in these sites. The truth is, these sites would not have come into existence without the involvement of many people and parties and such change only comes about with the constant evolvement of the combined action between local residents and visitors. We cannot just look at a picture that is packed with people and think, "Oh, this place is full of energy" or say "This place is lacking in vitality" simply because there isn't anybody in the image. There are many common factors that interact with each other that led to the continuous change a site goes through and this is something we can't determine by judging from the crowd of visitors or the number of people taking a stroll at the site.

Landscape design in Europe is different in the sense that landscape artist or architect assumes absolute power to decide what a site would look like in the future. Instead, the people and residents that live at the site would get together to discuss the strengths and weaknesses of the site. They would invite the landscape artist to participate in the discussion and voice their thoughts on their living environment in the future. The result of such discussion would determine the design of their site, rather than the thoughts and ideas of any individual landscape designer.

Wang: That would be what we commonly refer to as public participation. The truth is, the general public would be the users of open space, and landscape architects should not have the power to decide the appearance of landscape for a site. The message that Professor Bava wishes to express is that in the process of design, landscape architects must show respect for

吴雪萍（以下简称「吴」）：非常高兴能跟亨利巴瓦大师有面对面交流的机会，我想问几个比较具体的问题。我们西安的学校和法国波城农业技术学校进行合作，我们每年派4位学生到法国的学校去学习园林设计，法国也派4位学生到西安来学习。持续进行了两年，我自己比较疑惑的是，我们的学生跟着他们学习，最重要的是学什么，他们的学生来了以后，最重要的是我给他们讲什么？也就是说中国和法国的园林艺术，最重要的差异在什么地方，我们最应该补充的或者相互借鉴的在什么地方？

巴瓦：首先得把语言学好。无论法国学生来中国，或者中国学生去法国，最好的学习语言的方法，融入对方文化的方法，就是找一个男朋友或者女朋友（观众笑！）。

吴：我们的学生去了以后，在课程安排上，给他们学习什么样的东西？包括他们的学生来了以后，除了语言学习之外，我应该从哪几方面去安排他们的学习？

巴瓦：中国和法国两个国家的景观制度不一样，法国的甲方非常注重跟设计师之间的交流，并且甲方内部本身也会做很多的研究，在方案立意开始阶段就会做前期研究，这些工作都是甲方做的。我希望中国学生来法国学习，不只是到一些单纯的景观设计公司来学习做景观的方法，而是来了解法国景观整个流程和制度是怎么样建立起来的，包括设计师和景观之间的对话是怎么样建立起来的。

德国和法国的甲方是非常开明的，在巴黎西部的一个区域规划中，他们召集了所有城市的市长，因为这个规划是一个跨地区的规划，很多城市的市长在一起做个模型，就是说你作为一个当地的市长，你肯定最清楚你的居民需要什么？你作为一个市长最需要什么样的景观在你的场地里面。所有人在一起做一个模型，由他们来选择自己最需要的东西，再加以综合考虑，最后来做成一个大家都满意的景观设计。

唐：刚才说到法国景观园林设计制度和中国有些不一样，亨利巴瓦也提到过在设计当中经常用一些透水性水泥，以及法国的一些遗产保护和德国的不一样，或者说屋顶绿化拓展空间，还有场地的研究，公众参与等等。我比较关心的是，法国园林师、景观师的注册是怎么样拿到工程？这种制度是怎么样一个系统？公众参与有政府政策的项目？是国家的法律法规？还是作为一个设计公司，一个设计师的一种自觉的行为。

巴瓦：事实上在欧洲，特别是在法国，景观这个行业被当作一个专业的行业大约在七五年的时候，所以它在制度方面并没有那么健全。当然建筑领域会比较早一些，景观行业大部分还是跟着建筑这个系统来跑的。换句话说，他们现在还没有建立一套属于景观师的一个系统。

马一鸣：我在法国学习多年，对法国的教育比较了解。在法国景观教育确实有一套独特的体系，它不像国内的这些普通综合性大学，在法国，景观设计不在综合性大学当中而是在专业性大学设立，只有四所景观建筑学院能够授予法国国家景观设计师DPLG文凭，分别是凡尔赛，里尔，波尔多和马赛国立景观与建筑学院。学制是四年。入学要经过全国统一的考试，每年有120人能够进入这四所学院，同时有80多人获得文凭。其不光是一个法国国家颁发景观设计学的最高文凭，最重要的是法国注册景观设计师的资格证。

如果有这个资格证，你才能够在法国建立自己的公司，签署30万欧元以上的公共项目。如果没有这个文凭的话，你也可以被称作是景观设计师，但是没有这个签字权的，这个是有法律规定的。

王：我补充一下。亨利巴瓦教授刚刚有提到，法国在七五年以后才开始将景观当作一个专业领域，之前都是在园艺领域，学校的园艺学院包括现在凡尔赛国立园艺学院，后来才慢慢转成景观学院。

the target (and or potential) users. For example, in the design of residential landscape, if the architect intends to incorporate a green roof as a residential landscape, the process must involve the participation of the residents because rooftop garden spans across multiple households; while some residents would like the idea, others may feel otherwise. Landscape design isn't so much about "What I have to offer you"; it is about "Knowing what you want".

Bava: That is correct. The truth is, techniques and approaches to landscaping have gradually changed in France, Germany and other western European nations. Landscape architects have profound respect for the communities residing at the sites and the opinions that they have. These landscape architects would communicate with the residents and work together towards the project's completion, rather than having the landscape architect to make arbitrary and solitary decisions that will determine the course of landscape design.

Gao: I do have another question. In your afternoon presentation, you mentioned a project on a Seine River island that involved a rectangular park. I would like to know what basis was used to determine the format, area and location of the park.

Bava: This park, known as Billancourt and located in the west of Paris, is a rain garden. However, we were not the ones to decide the area or the format of the garden. Sometimes when we work on a major city project, we get to decide aspects of the design (such as the area of coverage), but that wasn't the case with this project in question, because it was a joint project that involved architects, city planners and landscape designers to conduct a research on the entire area. The architects and planners gave the definition for zone planning in the city and offered approximate format, location and area of the garden, and we received the assignment to design the garden. In France, landscape design is rather traditional: Party A commissions a mission to you, then you design it. Of course, such model of design is still quite different from China; in China; the commissioning party (Party A) would define details such as the area and format of the park.

Wang: Obviously, Professor Bava is emphasizing the importance of interaction between designers and users.

Yufan Zhu (Name as 'Zhu' below): At lunch today, I took the opportunity and asked Professor Bava a question, "As a European, what's your take on the United States?" He gave me a one-word reply, "Big." Personally, I would say that in comparison with Europe, America's culture has its problems. Because of teaching, I had many opportunities to visit Europe in the past two years and I managed to further my understanding of European cultures. Earlier this year I made a visit to Spain because of a collaboration and as I discussed cultural characteristics with other professors there, I made two distinctive observations.

I went to visit the studio of a renowned post-modern architect. His studio was renovated from a cement plant that was abandoned in 1972 and I think his works were similar to what Peter Latz after the 1920s. Next, I also went to see a landscape that won the award of best design for European Landscape Award; it

图08 林中居5
Fig08 Bois Habite5

他也提到，景观园林这一块，目前在法国还没有被保护，意思就是说，像建筑师因为被保护，所以才可以做建筑的设计。优点就是比较自由，没有法律的约束，也就没有法律上的责任。缺点是，没有受到保护，任何人都可以做。这个问题在美国也曾经有过争议，我有同学在那边做景观师，他们分两派说法，景观师要有注册或不要有注册制度，其实这攸关景观师想不想或要不要负法律责任的问题，美国在制度方面的发展已相当成熟了，每州有每州的注册制度，在台湾我们正朝着景观师执照制度的方向努力当中。

他还提到，正因如此，而给景观师一个很大的空间去自由发挥，但前提是你要找到一个好的业主。但是好的业主，并不表示一定要有钱，虽然钱非常重要，但是更重要的是他可以让你的想法落实。就某种程度来讲，你提出一些想法，他就在那个框架里提出他的概念，同样的道理，未来的城市发展不是有钱就可以把城市建起来。目前的状况是，在这个框架里，你可以很自由的找建筑师，找水土保持或者一些水利专业的人来跟你合作，在这样一个自由的情况之下，可以得到很好的设计或结果。

特别是对领导来讲，刚才提到的城市发展，可以从景观专业角度中提出一个愿景，根据这个愿景，可以明确将来如何发展，如何制订政策，制度就是从这个愿景开始的。

巴瓦：谢谢王教授，替我回答了大部分的问题！（观众笑！）

唐：亨利巴瓦是做设计公司的，设计公司就要顺应市场，我们易兰设计院也是如此。我很赞同他的一个理念，就是他提出有些项目是引导规划。在法国的项目中，可以由建筑师去引导建筑。我想知道，

was a memorial converted from a heritage site during the Franco regime. Seeing the landscape was a truly inspirational experience for me, because I have never imagined a memorial converted from an ancient castle could be that beautiful.

Bava: Well, when it comes to comparing the strengths and weaknesses of European landscapes and American landscapes, the task is still quite difficult. One of the more notable American landscapes that we know of would be a project that James worked on.

Zhu: The mastery of European landscape over site characteristics lies in explicit values. It could be an apparent trait that the designer has been focusing on and the designer would in turn unleash the potentials of said trait. The case study we just saw was outstanding, no doubt. But if we could achieve perfection, this dialogue with you would have been redundant. It is with certainty that I have discovered problems we had, and this why we need this exchange. What I would like to learn from Professor Bava is, during the process of value transition that Europe has undergone, how would you describe the transitional process, from the perspective of a European? I learned that European designers have come to place greater emphasis on the development of potential cultural values, which they would cautiously preserve. I would like to know, from Professor Bava's standpoint (just to clarify, I don't think Europe is perfect), his concept on this topic.

图09 林中居总平面图
Fig09 Bois Habite Master plan

建筑师在实践当中是怎么样去操作的？他的想法，执行力能够进行到什么程度？

巴瓦：现在在我的公司有建筑师的团队，也有景观师的团队。当今景观的概念慢慢趋向于城市化，但是建筑师又慢慢趋向于景观，这两方面渐渐靠近，我认为这是一个很好的现象。那谁来当头？谁来带领？我认为这要根据项目讨论的结果来决定应该往哪个方向发展。我们公司主要是做景观，所以就由景观师来领导，但是需要大家的共识，需要大家互相了解，互相认同，这在我们公司是非常重要的。

所以未来如何发展是由景观师来领导的。在我们公司，这个问题没有冲突，而是一个合作的状态。前面提到，建筑已经慢慢趋向于景观，跟过去的情况已经有了改变，景观也慢慢趋向于城市设计的方向。所以，这两个领域慢慢互相渗透，互相交流，互相合作。

我的公司组织除了一部分是景观师，一部分是建筑师，还有两到三位是环境师。我认为生态这个部分是非常专业的，当我们的项目牵扯到一些有关生态的问题，特别是针对场地的本质，也就是它的自然特性或者地质、地理等等跟生态有关的部分，我们就会找这方面的专业来跟他们一起协作，特别是提到德国项目的时候，因为德国对这方面非常重视，做德国项目的时候我们一定会找这方面专家跟他们一起合作，因为了解那个场地的特性是非常重要的，他们公司有生态和环境背景的专家可以去处理这样的问题，我们公司需要有这方面的专业背景的人才，来跟他们一起合作。

Bava: From the developmental history of European landscape, anyone from Europe would probably be impressed and fascinated by the sight of a beautiful landscape, even if he wasn't a landscape designer or someone familiar with the discipline. But the European landscape architects we see today emphasize more on the things we don't necessarily see. When we see a landscape that suffers from obvious problems or defects, we become more interested in it. During the presentation in the afternoon, I mentioned a quarry park situated between two German cities. Now, the French people feel that industrial historic sites are not worth preserving because France has demolished all such sites. In contrast, the Germans think an industrial historic site may be very beautiful and feel a sense of responsibility to preserve it. That is the difference between the culture of the two countries. Perhaps the French should learn from the Germans in this regard: there are things that deserved to be preserved.

Back in my firm in France, we never use expressions such as "beautiful" or "not beautiful"; we can't appraise an object to be beautiful or not beautiful. Studies and designs are conducted on the basis of special characteristics of sites, and one simply does not make a direct judgment on whether a site is beautiful or not. In this business, such arbitrary expressions are strictly forbidden.

唐：还有一个问题，我们都知道法国像中国一样很富有历史底蕴和文化内涵，法国也有很多的历史遗产，作为一个现代的设计师，您的公司参与了上海世博的法国馆的设计，我们易兰国际设计公司在上海世博做中国馆的设计，我想问一下，您在法国或者在欧洲园林设计，有没有试图创造现代欧式景观？

巴瓦：在做法国馆的设计和施工的时候，我们试图融入一些中国的风格，尽量避免法国的风格，可是到后来还是有法国痕迹。事实上对于法国，大家印象最深刻的就是非常不人性化，所有植物都去修剪，修剪成他想要的东西，那是时代的背景，我们从法国馆里还是看得到的。在世博照片里我们看到的是立体的画，可是它的平面还是有凡尔赛的味道。

我是从一个非常自由的思考去做我的设计，我的作品是几个合作伙伴头脑风暴出来的结果。一个团队里面的队员一定要有不同的专业背景，那样才能够激发出不同的火花出来，否则设计就会非常的干燥，没有营养。我们试图从不同背景去讨论同一个议题，让我们的作品更加丰富、有营养。我的公司在做法国馆的时候，是以这样的思维去做概念的设计。所以不要有所谓的风格，不要固定在某种风格，最重要是回归场地本身的特性。场地最初都是不毛之地，跟周边的城市做一些结合，让设计连接周边城市的特性，而不要刻意去营造某种风格。

一定要尊重场地。即使试图把场地转化成另外一个东西，但是场地本身的精神还在。接下来跟朱院长谈到你的作品，说有一个湖，这个我就不知道了，您来说吧！

朱：总而言之，亨利巴瓦强调的是尊重场地，还有一个问题针对场地的特殊性的。当你认为某些不存在的价值是有价值的时候，你会完全改变你的世界观。

王：接下来，在座的贵宾有没有人要提问？

提问1：我是来自海南大生态园林景观的，海南是中国唯一一个国际旅游岛，地处于中国亚热带和热带，湛蓝的大海，洁白的沙滩，清新的空气，每当我们到海边就感觉到非常开阔，是一个很好的地方。可是海南的城市设计都是东抄西抄，东南亚风格也好，巴利岛风格也好，地中海风格也好，唯独没有自己独特的园林景观风格，今天很荣幸来到现场跟各位大师和各位同仁，能够一起探讨这个问题，我希望在座的大师和同行，能够给咱们的宝岛提一个建议，海南的设计规划要往哪个方向走，怎么走才能做出一个独特的带有国际旅游岛性质的园林城市。今天我把这个希望带给在座的各位。希望咱们的宝岛，以后比巴利岛会更好。谢谢。

巴瓦：欢迎您和我一起合作，一起创作出新的东西。

提问2：怎么样来看待景观弹性的问题？

巴瓦：人口的增加，使得气候发生变化，这是全球性的问题。密度增加跟生物多样性的增加应当要同步进行，但是目前人口越来越多，相对会让你的生物多样性越来越少。所以我们的目标就是要试图把这两者拉近，换句话说，当我们在做一些建筑或者人工地毯的时候，我会试图去增加很多绿地和栖地，让野生动植物能够在里面自然成长。

另外，我们可以利用人工的方式来弥补因密度增加所造成的环境问题，比如说雨水的回收，同时增加排水系统、集水系统等等。我们公司现在正在试图朝着这方面跟相关的专业来合作，希望我们团队可以做出一个更好的结果。我们公司不只是做景观的设计或者施工，我们还有研究员，研究员就是针对他刚刚讲的这些特殊的问题去做深入的研究，研究的结果用来给我们去做规划设计。所以我们公司不是只有规划设计，还有背后的专业知识，这些专业知识是从研究人员那里得来的。

图10 林中居6
Fig10 Bois Habite6

Prior to the 1980s, most French and German designs involved the identification of an original form; for example, French or English gardens to be applied to a site. But in the end, we would find characteristics about the site that were worth further exploration so we couldn't just simply apply the form to the sites, we had to study them.

Before the 1980s, the value of European landscape works surpassed the value of sites themselves. And as such, we focused on creating works that would be of higher value compared to the site. But after the 1980s, we realized that the value of the sites is actually more precious than the value of landscape projects, so we wanted to dig deeper into the intrinsic qualities of sites. We knew that when we could flawlessly combine sites and landscape projects together, the result would be perfection. And that is what we are after.

Wang: That was an excellent question from Professor Zhu. We should recognize the fact that different European countries have their corresponding uniqueness. When we refer to "Europe" today, we are in fact talking about a broad scope that covers a significant territory. Although European nations today are all

图11 比扬谷公园1
Fig11 Boulogne1

景观团队一定要跟研究团队密切合作。研究者有很高度的创造力，对他们来讲，他们所研究出来的东西很有创新性，而这些新的东西可以注入到我们的项目里，这是一个很好的资源。但是毕竟我们还是人类，我们现在所做的东西还是为了人类能够有一个好的居住环境，一个健康的生活环境。当然这个健康不单单是指人类，整个环境是达到一个生态平衡的状态，这才是人类居住最好的环境，我们要的东西不是量，而是质。

接下来再补充一点，我们应当从错误里面去得到正确的做法，或者是比较合理的做法。在七０年代，他们犯了一个很大的错误，从那个时候起他们就觉得应该从错误当中去学习。七０年代的房屋是连栋式的，高层的，多居户的房屋，在法国巴黎，他们的房屋比较低矮，一般是4-5层，独户一栋的，他们的房屋跟中国的居住区十层左右高的楼房类似。他们在七０年代的时候大量构建了这种他们所谓的高层建筑，而且是连栋的高层建筑，使得整个环境都受到很大的破坏，造成了都市的品质的下降。

提问3： 亨利巴瓦教授和各位嘉宾的交流让我很有感触，特别是刚才讲到对场地的解读的时候，提及欧洲场地的价值要高于设计师的价值，我觉得这是一个新方向。在我的理解中，场地解读是一个社会最基本的工作思路，解读场地实际上有两种方向，一个是通过场地更好来体现设计师的思想，一个是真正保留场地的价值。我想问一下亨利巴瓦教授，在你的理解上，场地价值具体分为哪几类，是哪些内容？

巴瓦： 一个场地的价值是已经存在的，它是一个历史，身为一个景观师必须走入那个历史里面跟它对话，换句话说，你要跟土地去对

members of the European Union, they still retain their own cultural features. In response to Professor Zhu's question, Professor Bava has learnt from the German people to respect issues that others may have overlooked and to try and improve upon it when the German government perceived value in the preservation of an abandoned mine. Professor Bava also mentioned that it was something that has never been emphasized previously in France; a specific kind of cultural characteristic. Professor Bava has been involved in numerous projects all over the world and given his profound experience and exposure to the world, I am confident that he has a very good grasp on the knowledge he still needs to learn and what he could learn from other countries. Through such international exchanges, he would be able to share his experiences and insight.

Xueping Wu (Name as 'Wu' below): It is such a pleasure for me to have this opportunity to speak to Professor Henri Bava face to face. I do have a few specific questions that I would like to ask. The Xian landscape Industrial School and LEGTA de Pau have been involved in an exchange program whereby we would send four students from China to LEGTA de Pau to study landscape design and LEGTA de Pau would send four of their students to China for the same purpose. This exchange program has been going on for two years and this is what I have been wondering: What would be the most important thing that our students should

话。更重要的是，我认为一个景观师必须要听这个场地说话，用一个非常法国式，非常罗曼蒂克的说法就是，要慢慢去体会，慢慢去理解，这跟当时设计罗浮宫是有点像的，设计之前去三、五次，去感受当中的环境。

回到刚刚讲的问题，一个设计师一定要去听，去认同，去了解，到底土地在跟你说什么，你不能跟他大声说我要你怎么样，你应该怎么样，而是你应该慢慢的停下脚步去听他，这是我的逻辑，我的哲学思维。场地的价值是什么？其实价值本来就在那里，你要去对话才能知道。

就时间上来讲，我们只是去做短暂的设计，而那个场所是长期就存在的，意思就是我们是渺小的，我们要去尊重场地，要尊重自然，而不是我们要去克服它。我们要从错误里面去学习。

王： 亨利巴瓦所提到的设计理念，跟我们从历史上对法国刻板的印象不同，他是凡尔赛设计学院毕业的，这是法国最高的园林学院，他的设计理念跟我们对传统法国庭园非常强制性的去把这个场地给做到我要你怎么样就怎么样的那种哲学不同，他非常尊重这个场地，我们做为一个设计师要去尊重场地。你为谁设计？使用者是谁？场地里面就是因为有人才有活力，才有生命，不是你自己要去强加给他们的，所以他是一位很谦卑的景观师，他尊重场地，尊重使用者，强调的是民众参与；尊重历史，尊重场地，从这里切入，来作为设计的出发点。

由于时间的关系，各位来宾如还有问题，请在会后再向亨利巴瓦教授请教。感谢各位嘉宾献出宝贵的时间，一起和亨利巴瓦教授对谈和分享。谢谢各位。

---

learn over in France? When LEGTA de Pau students come to China, what knowledge should I impart to them? In other words, what would be the most important difference between the landscape arts of China and France? What should we be learning from one another or what should we do to complement one another?

Bava: First you must become proficient in the language. It is the same thing for French students studying in China or Chinese students studying in France; the best way to learn a language and to blend into the local culture is to find yourself a boyfriend or girlfriend! (laughter from the audience)

Wu: What can our students expect to learn, in terms of their syllabus, when they go to France? I guess this question goes both ways. When students from LEGTA de Pau come to China, what kind of arrangement should I be preparing (apart from language lessons) to ensure effective learning for them?

Bava: Well, obviously China and France have different landscape systems. Party A (competent government authority) in France places great emphasis on the interaction with designers. In fact, Party A would conduct many internal studies and get started with preliminary research at the initial phase of project initiation. These are tasks that Party A would undertake. I hope that when Chinese students visit France to study, they would not limit themselves to just going to some traditional landscape design companies to learn landscape techniques. Instead, I would encourage them to take the initiative to find out how the entire process and system of French landscape was established, including how designers and landscapes engage in dialogues.

Party A in Germany and France are very open. In a zone planning operation implemented in western Paris, the government summoned all the mayors of relevant cities for the effort because it was a cross-territorial planning. The mayors were gathered together to build a model. The idea is that, as mayor of a city, you would have the best knowledge as to what your residents want; as mayor, you would know the kind of landscape you need in most in your city. So everyone would get together to build a model, and each participating mayor would choose the item they needed most. All their selections would be taken into consideration to create a landscape design that would satisfy everyone.

Yanhong Tang (Name as 'Tang' below): It is interesting that you brought up the fact that France has a different landscape and garden design system from China. You have previously mentioned that you often use water-permeable concrete in your designs and that France differs from Germany in terms of approaches to heritage site preservation, rooftop greening expansions, site researches, public participations and so forth. Personally, I am more curious about how licensed landscape designers in France get their projects. What kind of system are we talking about here? Does it involve special projects that include public participation in governmental policies? Or is it state regulations? Or is it simply an act of self-awareness for a designing firm/designer?

Bava: In reality, especially in France, landscape has only come to be perceived as a professional line of work at around 1975. Naturally, the system that governs the business is less than comprehensive. Of course, the construction industry has received its due recognition earlier and the majority of the landscape sector still adheres to the system we have for the construction business. In other words, we have yet to create a system that is exclusive to landscape architects.

Yiming Ma: I am a student who studied in France so I have a fair understanding of the French education system. In France, landscape education does have an unique system and it is nothing like a syllabus offered at traditional general universities in China. In France, you won't find landscape design courses at general universities; such courses are only offered at technical institutes (such as Ecole Nationale Supérieure du Paysage de Versailles) or schools of architecture and landscape. Such schools are stronger in terms of their specialization. The school I studied at only offers two majors to choose from: landscape and architecture. Students must finish four years of training and only those who have completed two years' worth of related subjects at a French university would be eligible to take the enrollment test. Finally, you must take a standardized national test to determine if you would be admitted to the school. The diploma that you receive after four years of study is more than just a doctor phase.; it also comes with the most crucial credential of being a qualified landscape architect.

This credential is the pre-requisite condition to establishing your own landscape design firm in France or landing a project worth over 300,000 Euro. You can still take the title of "landscape architect" even if you don't have the diploma, but you would not be entitled to accept projects worth over 300,000 Euro. This is clearly stated by law.

Wang: Allow me to add to the point. As Professor Bava has just pointed out, landscaping was only recognized as a professional discipline in France after 1975. Prior to that, it was perceived as a part of horticulture schools specializing in the discipline (including Ecole Nationale Supérieure du Paysage de Versailles) were known as horticulture schools. Over time, they gradually became known as landscape schools.

Professor Bava also commented that the discipline of landscaping has yet to be protected in France. What he is trying to express is that architects are "protected" by their privileges and credentials that enable them to involve themselves in the design of buildings. The upside of not having a state-established system that governs the sector is that since there isn't any legal regulation, one would never be held legally liable for his work. However, the downside of not having such a system is that practically anybody could claim to be a landscape architect. In fact, this issue had its fair share of conflicts in the United States as what he knew, where I have a schoolmate working as a landscape artist. In the United States, views on this issue are divided between two sides: One side opts for the need of an official body to govern the licensing of landscape architects while the other side opts against the idea. If you think about it, the issue can be interpreted as "Should or do landscape architects want to assume legal liabilities?" Now, we all know that when it comes to such systems, the US has developed a very mature system whereby each state has its specific licensing system. In Taiwan, we are currently working towards establishing a permit system for landscape architects.

Professor Bava also noted that, because of this, landscape architects in France have a lot of room to utilize their designs, under the premise that they find a good proprietor. It is important to note that a good proprietor doesn't have to be a rich one; although money matters a lot, it is more important that your proprietor gives you the freedom to materialize your ideas. To put it in another way, a good proprietor is someone who will share his concept with you in the framework of ideas that you throw at him. The same thing could be said for urban development – it will definitely take more than just money to achieve ideal urban development in the future. The current situation is, under such framework, where one can freely enlist the help of architects, experts in soil and water conservation/hydrology for collaboration. Such freedom usually leads to great designs and results.

Chinese government officials can also refer to the idea of urban development we have just discussed and propose their visions from the professional perspective of landscaping. Based on such vision, it would offer a clear guideline for future development and formulation of relevant policies. In fact, vision is starting point of systems.

Bava: Thank you, Professor Wang, for answering most of that question on my behalf! (laughter from the audience!)

Tang: Professor Bava runs a design company and design companies should follow market demands. And it is the same with ECOLAND Planning & Design Co., Ltd. I fully agree with an idea that Professor Bava talked about – projects that involve planning guidance. With French projects, architects have the freedom to lead construction projects. I would like to know, how does an architect go about achieving this through practice? How far could the power of execution carry his ideas?

Bava: Now my company has a team of architects and a team of landscape designers. The concept of landscape today has gradually shifted towards urbanization, but architects have gradually shifted towards landscapes. So we have both sides slowly coming together, which I think is a good thing. But who will be the head? Who will take the lead? I think the direction of development should be determined based on the result of project discussion. Our company specializes in landscaping, so landscape architects would lead our projects. But it will always take consensus, mutual understanding and mutual recognition; these are of extreme importance at our company.

In other words, our future development will be spearheaded by landscape architects. In our company, this is not a conflicting issue but a state of cooperation. As I have previously stated, architecture has gradually shifted towards landscaping; the situation has changed and landscape has also slowly inclined towards urban design. And so, the two disciplines will slowly to permeate one another for mutual interaction and collaboration.

Apart from the landscape designers and architects at my company, the organization also has a few environmentalists. Personally, I perceive the aspect of ecology to be highly professional. Whenever our projects involve issues relating to ecology, especially factors pertaining to site characteristics (i.e., the natural characteristic or geological/geographical/ecological traits), we would seek experts in this area and work with them. This is especially true with projects that involve Germany, because Germany places great emphasis in this area. So when we work on a German project, we would always collaborate with experts in this area because understanding site characteristics is very important. Their company would have experts with background in ecology and environment to tackle issues like these and our company needs talents who specialize in these areas for collaboration.

Tang: I have another question. We all know that just like China, France is also very rich in historical and cultural background, not to mention numerous historical heritage sites. As a contemporary designer, you and your company took part in the design of the France Pavilion in Expo 2010 Shanghai China (ECOLAND was commissioned to work on the design for the China Pavilion). Here's my question: have you ever tried to create modern European landscape in your works of French or European garden design?

Bava: In the design and construction of the France Pavilion, we tried to incorporate some Chinese elements and avoided conventional French styles. But in the end, you could still notice signs of "Frenchness" in the work. In reality, the most dominant impression that people associate with France is how everything is far from being human-centered; we trim all our plants into the shapes we want. That is the background of our time and we can still see that from the France Pavilion. Although we see a 3D image from pictures taken at Expo 2010, you could still pick up hints of Versailles from its graphic illustration.

Personally, I work on my designs from a very liberal and

图 12 比扬谷公园 2
Fig12 Boulogne2

unrestricted mindset; my works are the results of brainstorming sessions I had with my business partners. It is important for a team to comprise members of different professional backgrounds in order to generate different ideas and trends of thinking. If not, the design would be very dry and malnourished. We always attempt to discuss an issue from different backgrounds in order to enrich and improve upon our works. When my company was working on the France Pavilion, we went about the conceptual design with that mindset. In other words, don't be focused too much on the so-called "styles" or be limited to any one of them; the most important thing is to return to the characteristics of the site itself. All sites are nothing more than barren land at the beginning; we take them and connect them to the neighboring cities through designs that link to the characteristics of neighboring cities, instead of intentionally creating a style.

One must always respect the site. Even if you attempt to convert the site into another object, the spirit of the site will still remain. I was talking to Professor Zhu regarding a work of yours, I think it was a lake. I am not too sure about this; why don't you take over?

Zhu: Summing up what you have just said, Professor Bava has emphasized on the respect for sites, and a site-specific question. I would just like to add that when you perceive value in something that others see as worthless, your worldviews will be changed completely.

Wang: And now we will have our floor session. Is there anyone from the audience who would like to raise any questions?

Question 1: I am from Hainan Ecological Landscape. As you know, Hainan is China's only island of international tourism. Located in-between the tropical and subtropical regions of China, it is blessed with great blue sea, white sandy beach and clean air. Whenever we go to the seaside, we feel reinvigorated because it is such a nice place. However, the urban design of Hainan is merely a collection of imitation, be it Southeast Asian style, Bali style or Mediterranean style. There is no such thing as a unique landscape style that Hainan can claim to be its own. Today I am deeply honored to be here to discuss this issue with Professor Bava and other distinguished guests. I am hoping that everyone here could offer their insight and suggestion as to where Hainan's design and planning should be headed in order to turn the island into a garden city that has its exclusive qualities for international tourism. I come here today with this hope that our precious island would one day surpass Bali island. Thank you!

图 13 比扬谷公园 3
Fig13 Boulogne3

Bava: I'd love to work with you and hopefully create something new together.

Question 2: What is your view on the topic of landscape versatility?

Bava: The growth of the population around the world has led to the climate changes we have seen in recent years is a global issue. The increase in population density and biodiversity should be in sync, but what we have now is a rapidly growing population that causes biodiversity to decline. So our goal is to try to bridge the gap between the two. In other words, when we work on an architectural product or lawn, I will try to add more greenery and habitats so that wild animals and plants could grow naturally.

In addition, we could use artificial means to make up for the environmental problems caused by the increase in the population density. For example, the collection and use of rainwater, coupled with drainage system, catchment system and so forth. Our company is currently trying to collaborate with experts in this area in the hopes of delivering better results. The operation of our company is not limited to just the design and implementation of landscape; we also have researchers who are responsible for carrying out in-depth studies on the special topics that we have just talked about and the results would serve as useful references in relevant planning and design. In short, our company does more than just planning and design; we also have relevant professional knowledge that came from the hard work of our researchers as our backup.

It is crucial for the landscape team to work closely with the research team. Researchers are highly creative, and as far as they are concerned, the results of their studies are very innovative. And such novel ideas could be incorporated into our projects as an essential resource. Nonetheless, we are just human beings after all; we are still driven by the objective of creating an ideal environment for mankind to live in; a living environment that is healthy. Of course, the concept of "health" is not limited to just human beings; it also involves helping the entire natural environment to achieve the state of ecological balance; that would be the best environment that human beings can hope to live in. What we want is not quantity, but quality.

I do have one more point to add that we should always learn from our mistakes and try to do things the right (or in a more logical) way. In the 1970s, they made a huge mistake and from then on, they have come to realize that something must be learnt from the error. In the 1970s, high-rise, multi-household townhouses were the dominant type of housing. In Paris, houses were generally shorter than that; a typical building would be 4 or 5 stories high and would accommodate only one household per building. In short, their houses are quite similar to the 10-story houses you can find in China's residential areas. During the 1970s, they built a massive amount of these so-called high-rise buildings that were connected together like townhouses. Consequently, the environment sustained significant damages and led to the decline of city quality.

Question 3: It has been truly inspiring to witness the exchange between Professor Bava and the guests, especially Professor Bava's interpretation of sites and his remarks that the value of sites (in Europe) surpasses that of the designers. I think that is a new direction that we could take. In my understanding, the interpretation of site should be the most fundamental working logic for any society and it may be achieved in two ways: 1. To achieve higher degree of materialization for designers' ideas through the site and 2. To truly preserve a site's value. My question for Professor Bava is this: Based on your understanding, how would you categorize site values specifically and what contents would you associate with them?

Bava: The value of any site must be pre-existing. It is a history. As a landscape architect, you must go into that history and communicate with it. In other words, you must have a conversation with the land itself. More importantly, I believe a good landscape architect must be willing to listen to what the site has to say. To put it in a very French, romantic way, you have to slowly appreciate and understand the site. The concept is quite similar to the designing process of the Louvre; prior to the design, the designer had made several visits to the site to feel the environment.

Returning to the question you have just asked, a designer must listen, acknowledge, and understand what the land is trying to tell you. You can't just yell at the land and say, "Here's what I want you to do," or "Here's what you should do." Instead, you must slow down and try to listen to the site. That's my logic and philosophy. What is the value of a site? The truth is, it is always there for you to find and the only way you will know it is to have a dialogue with the site.

From the perspective of time, what we do is temporary designs for sites that last for a very long time. In other words, we are very trivial and insignificant in comparison. We must learn to respect sites and respect nature, instead of wanting to conquer it. We must learn from the mistakes we have made.

Wang: Professor Bava's design philosophy is different from the stereotypical French design styles that we have come to be familiar with from history. He is a graduate of Ecole Nationale Supérieure du Paysage de Versailles, the most prestigious school of landscape and gardening in France. His design concept is very different from the "I want to turn the site into this and that" coercive philosophy we found in traditional France garden and he has great respect for all sites. As designers, we must learn to respect the sites that we work with. We should always remind ourselves who we are creating the design for and who the users are. A site finds its energy and vitality from the people present at the site, and it is not something that you can forcefully give to a site. Professor Bava is a humble landscape architect who not only respect sites but also their users. He emphasizes public participation, respect for history and respect for site as the point of interception for his designs.

Due to time constraints, I would like to ask our guests to kindly keep any other question they may have till afterwards for Professor Bava. I would also like to thank all of our guests and the audience for taking the time to take part in this dialogue and sharing with Professor Bava. Thank you all!■

大师报告会对话嘉宾简介：
Dialogues with the guests at the presentation:

嘉宾主持：王小璘 教授／《世界园林》总编
Guest: Xiaolin Wang Professor / Editor-in-Chief of Worldscape

嘉 宾：高翅 教授／华中农业大学副校长
Guest: Chi Gao Professor / Vice President of Huazhong Agricultural University

嘉 宾：吴雪萍／西安市市容园林局副局长
Guest: Xueping Wu / Deputy director of Xi'an city bureau of parks and woods

嘉 宾：朱育帆 教授／清华大学建筑学院景观学系副主任
Guest: Yufan Zhu Professor / Deputy head of Landscape Architecture Department

嘉 宾：唐艳红／易兰国际设计公司副总裁
Guest: Yanhong Tang / Vice President of Ecoland Co., LTD.

翻 译：马一鸣／法国国家景观师 DPLG，法国景观协会会员
Translator: Yiming Ma / Landscape architect DPLG (National Diploma), member of the french federation of landscape

大师报告会及嘉宾对话时间：2012 年 4 月 27 日
地点：北京新大都饭店国际会议中心
Date: 27 April, 2012
Place: International Conference Hal of Beijing Capital Xindadu Hotel

杨育庭（中译）
Translated by Yuting Yang
何友锋（校订）
Chinese reviewed by Youfeng He

论坡地住宅小区之景观意象及塑造
THE FORMING OF LANDSCAPE IMAGE FOR HILLSIDE RESIDENTIAL COMMUNITIES

王小璘　何友锋　　　　Xiaolin Wang　Youfeng He

前言

城市人口不断增加，住宅用地无法满足住房的要求，坡地住宅成了未来的发展趋势。坡地环境由于具有起伏的山形、多变的线条、自然而调和的色彩、特殊的气候环境、新鲜的空气、产业景观、较少的人为活动和宁静的氛围，使其景观意象与都市迥然不同。坡地环境因其相对高差变化较大，一般除人为因素所造成的产业道路或登山步道之外，亦有部份因自然因素影响其景观。诸如地形变化造成谷间自然通路或山脊线，不同植群生长变化形成空间通路等。虽然其环境感知与一般道路并无差异，唯就整体景观意象而言，坡地道路则具有环境的主导地位和多样的自然特质。由于坡地视觉环境为地形、植物或水体等自然元素所组成，其视觉尺度相对较大，又因地形阻隔及人工道路较少，使坡地场所意象不易形成。纵有少数特殊组成的自然特征景物，亦因其位于连续性的区域中，使感知的机能性与活动性减弱，必需藉助人为作用塑造坡地住宅小区之景观意象。

坡地环境蕴涵之景观意象

景观是人类通过知觉官能和环境建立的关系，在此关系中，人是主体，环境是客体，人类经由视觉物理及生理、心理的调节作用和意识层面的思想运作，对环境产生不同的感知。景观意象即人在错综复杂环境中所感知和意识到的有组织的形象，它是一种对环境的整体印象，由观察者和环境交互作用所构成，并为人所知觉而产生的共同形象。景观意象具有两个要素：

(1) 观景者对所观环境产生的结果；
(2) 意象和意象连系时所产生的效果。

通过这一交互作用所产生的共同意识，形成人与人之间沟通的符号元素和共同记忆，不仅有助于人们对环境的认识，亦提供了人与人的交流机会和认同感。

坡地环境因其强烈的自然地域特性，而蕴涵着特殊的景观组成和视觉元素，其景观空间意象的产生，系由人为活动与地形、植栽、水域及气候变化等景观元素间的互动关系所形成，进而透过质感、色彩与造型及其组合的相对层次而呈现，其组织常与心理意象、知觉活动与实质形式相应和，且其组合型式亦因实质环境和知觉环境之不同而异。因此，坡地和住宅小区景观意象可由以下三方面来探讨：

1. 意象景观

意象景观系在实质形象的组成上，通过人类视觉心理的调节作用，重新组织所构成的心理意象景观。

坡地环境的组成元素包括丘陵、台地、草原、森林、湖泊、农田、果园、牧场或人工开辟的游乐园等，其所形成的视觉意象具有强烈自明性和连续性。就人类居住环境的感知而言，坡地住宅小区景观意象常因其组成元素较为复杂，而穿梭其间不易辨认方位。同时，由于坡地住宅小区环境是由土地经由许多片断的意象元素，如地形空间、形态、细部、象征、样式、用途、活动及居民等所组合，并向平面伸展，观者可自由出入其间，并由各区域之共同特质组成环境中独立且统一的体系，故住宅小区环境意象的强烈与否，须视内部组成所聚合的主题特色程度而定。

因此，坡地住宅小区意象景观之塑造应侧重形式的美感，以建立整体结构及塑造独特元素为目标，强化景观元素的组织与涵构为目标，

Introduction

Due to the increase in urban populations, the capacity of residential sites no longer satisfies the demands for residential development. Therefore, there has been an inevitable trend toward hillside residential developments. The hillside landscapes should present different images from the urban sites for its fluctuating mountain shapes, different skylines, natural colors, special climates, fresh air, less human activity and peaceful atmospheres. For example, besides those industry pathways or mountain hike pathways built for dramatic relevant height differences, there are also some pathways such as volley passages or ridgeline passages formed by topographic changes and interval passages formed by growth changes of different groups of plants. Though the environmental perceptions of such passages remain the same as general roads, in the view of the entire landscape image, hillside pathways are dominated by the variety of natural features. Due to the visual perceptions of the hillside environment being composed of natural elements such as topography, plants and bodies of water, the visual scales tend to be relatively big. Moreover, because of the lack of topographic segregations and built roads, it is hard to form senses of place for hillside sites. Though there may be few natural remarkable features, however, the function and mobility of the perception are still weakened by being in the continuous area. Therefore, such perceptions have to be strengthened and formed with the man-made images of hillside residential community landscape design.

The landscape image of hillside environment

Landscape is a bond relationship built between humankind and the environment with human sensations. In such bond relationship, humankind is the subject, while the environment the object. In general, humanity builds up different perceptions toward the environment through visual, physical and mental mechanism as well as ideological operation on the conscious level. The image of landscape, therefore, refers to the organized image that a person feels and senses in the complex environment. It is a whole impression composed of the interactions between the viewer and the environment. Therefore, it comes with two important elements: (1) the viewer's perspectives toward the observed environment and (2) the effects resulting from images connecting with images. With the common concept caused by such an interaction, the communication symbols and common memories are then established. It not only helps humanity knowing more about the environment but also provides opportunity for interaction and sense of identification.

Due to its strong natural geographical characteristics, the hillside environment therefore contains special landscape

专题文章 ARTICLES

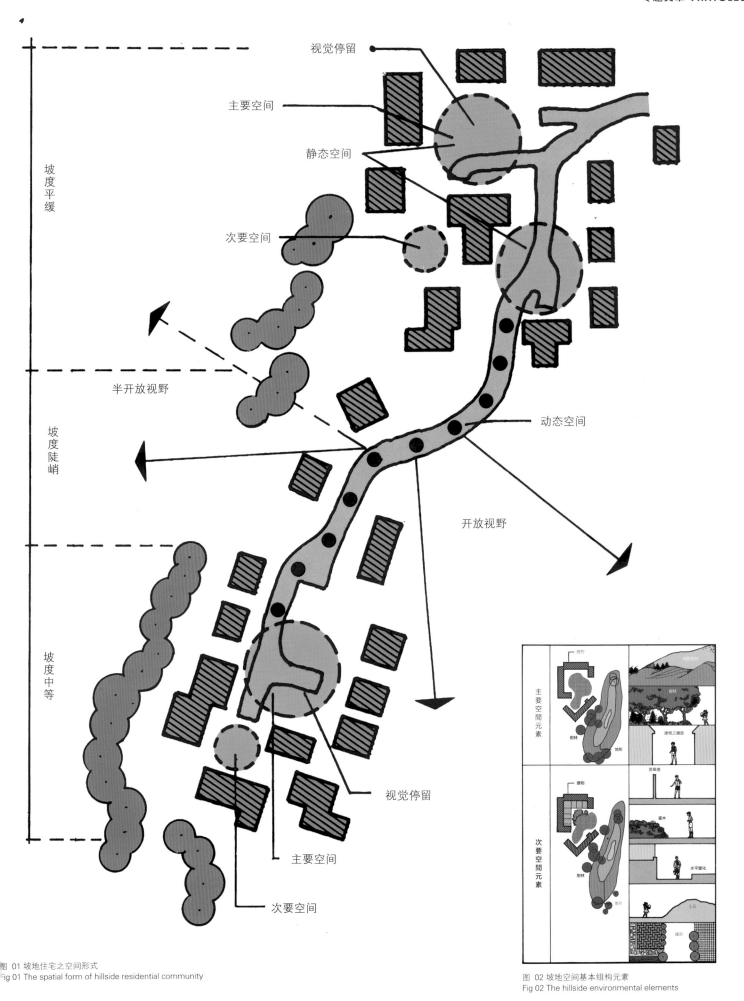

图 01 坡地住宅之空间形式
Fig 01 The spatial form of hillside residential community

图 02 坡地空间基本组构元素
Fig 02 The hillside environmental elements

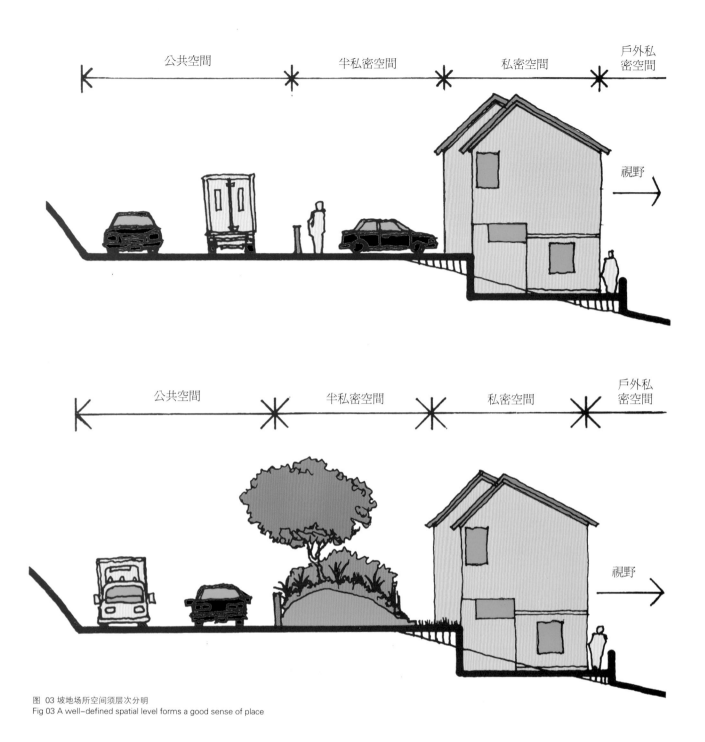

图 03 坡地场所空间须层次分明
Fig 03 A well-defined spatial level forms a good sense of place

并以坡地住宅小区整体尺度为设计对象，使全区景观结构明朗而清晰（图 01-02）。

2. 知觉景观

知觉景观系通过人类对景观实质元素刺激所产生而赋于特殊情感的空间形态。其景观空间层次已超出形式范围，必须藉由人群的聚集和活动而呈现特殊的面貌以构成环境知觉的形象。

坡地环境由于自然组成元素尺度大、景物绵延不绝，在视觉空间感知上极易造成强烈之联系意象及空间界定作用。其视觉边界的形成大都来源为自然景观元素，在视觉及实质上往往具有可穿越的特性。因此，坡地住宅小区知觉景观应侧重气氛和特色的营造，以建构场所尺度为目标，知觉景观塑造为目标，小区局部地区为对象，并以独特风格的知觉景观场所为景观塑造之重点，通过线性元素如道路和山脊线等，与整体景观结构产生有意义的内在关联。一个良好场所的塑造，须为空间场所层次分明，并具有方向的自明性和良好的视觉特征（图 03）。

components and visual elements. The landscape image results from the interactions among human activities, topography, plants, water fields, and climate changes; and it is presented in relevant levels as textures, colors, forms and their compositions. Furthermore, it has to correspond to mental images, perspective activities and physical forms, whereas the compositions may differ based on the actual environments and perspective environments. Therefore, the image of landscape design for hillside residential community can be discussed in the following three ways:

1. Imagery Landscape

Imagery Landscape refers to the mental image landscape reconstructed by humankind's visual and mental mechanism of physical images.

Hillside environment is composed of the elements such as

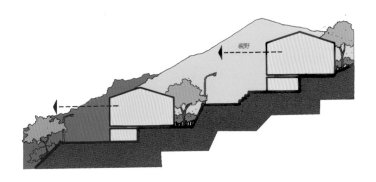

图 04 坡地建筑形式和布局与背景相互协调
Fig 04 The composition of form and layout creates the harmony for the community landscape

3. 实质景观

实质景观系通过景观元素间的组合关系，与实质单元及其细部处理所共同建立的视觉组合，主要在于强调景观涵构的功能及形式美感，为意象景观及知觉景观之具体呈现。

坡地实质意象的造型多以自然元素的形式、质感、材料、尺度和造型为主；其所组成的综合意象与周围景物调和，但常与地域或植被形成对比和清晰的剪影。故就环境感知而言，坡地实质意象的形成系通过细部的处理，在视觉元素相对较大和视线不易穿透的坡地环境中，塑造统一的调和而获致独特的坡地整体意象。

因此，坡地住宅小区的实质景观应侧重意象的强度，以形式与背景的和谐为目标，塑造细部尺度为目标，景观单元及其组合为对象，以具有独特的外在形式、质材、颜色、比例和尺度之差异与环境产生对比效果，以达致小区景观之协调性与生动性（图04）。

综上所述，坡地景观意象为一综合现象，其构成之差异乃一体三面。其中，知觉景观受实质景观组成形态的影响，同时也是意象景观形成的基础；实质景观经由视觉及心理的调理与组织，形成象征性的图示，进而构成意象景观。实质景观和知觉景观乃住宅小区意象景观的基础，无论意象景观或知觉景观之塑造，均须仰赖实质形象表现，故其设计应尽可能具体化和实质化，以满足其实用功能及视觉的美感需求，创造美好的小区生活环境。

其次，坡地住宅小区环境意象明显，线性元素自然性低，边界意象强烈，实质元素特性突出，故景观价值较高。加之坡地环境环境视

hills, tablelands, grasslands, forests, lakes, and farms, orchards, grazing lands or built entertainment parks with visual images of strong sense of identity and continuity. However, in the aspect of the sense of the living environment, living in such environment may result in lack sense of directions for the relatively more complicated elements of the landscape image for hillside residential community. In the meantime, the environment of hillside residential community is composed of many discontinuous image elements such as spaces, forms, details, symbols, patterns, utilities, activities and topography which extend horizontally; where the viewers are free to move around in the space. There are individual and united systems in the areas that share the same special features; in consequence, the impression of the community environmental images depends on the theme features united inside of the community.

Therefore, the goal in forming the imagery landscape for hillside residential community should focus on the establishment of the aesthetic perception by the entire structure and form the unique feature element in order to strengthen the structure and context of the environmental elements. The landscape structure of the entire area is made clear by focusing on the design and complying with the scale of hillside residential community (Fig 01- 02) .

2. Perceptional Landscape

Perceptional Landscape refers to the spatial morphology with special affections stimulated by physical landscape elements. Its landscape spatial level has been beyond the spectrum of images, whereas the perceptional landscape has to be composed of humankind's gatherings and activities.

Due to the large scale of the natural composition elements and the continuous views of hillside environment, it is easy to create strong impacts on relating images and spatial definitions for visual perceptions. The visual borders are mainly formed by natural landscape elements, which usually with visual and physical penetration ability. Therefore, the perceptional landscape for hillside residential communities should focus on the construction of an environmental atmosphere by giving with the aim being to constructing spatial scales, the goal to form perceptional landscape, the target as s small partial area. The emphasis is to create a unique perceptional landscape by creating meaningful interactions with the entire landscape structure, applying linear elements such as pathways and ridgelines. Therefore, it takes a well-defined spatial level, oriented self-identification and good visual features to form a good sense of place (Fig 03).

3. Physical Landscape

Physical Landscape refers to the visual combination constructed with the composition of landscape elements, physical units and the details. It is mainly focused on the functions and beauty of the landscape context as well as the presentation of imagery landscape and perceptional landscape.

The physical hillside landscape is mainly structured with the typology, textures, materials, scales and forms of the natural element, which are usually coordinated with the integrated images and surroundings yet contrast to the fields and vegetation. Therefore, in the aspect of environmental senses, it takes detailed treatments to form coordinated and unique hillside landscape images given with relatively large visual elements and difficulty of visual penetrations.

Therefore, the physical landscape for hillside residential

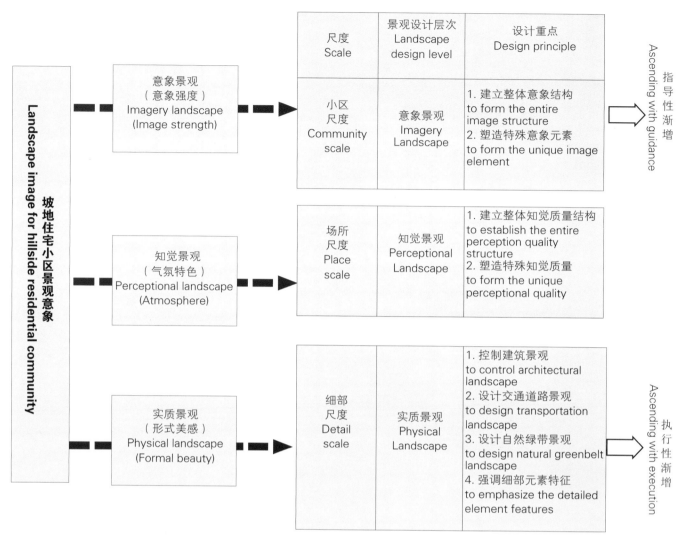

图 05 坡地住宅小区景观意象设计
Fig 05 Imagery landscape design for hillside residential community

觉元素尺度相对较大，视野的连续性及可穿越性较差，因此，如何保留坡地起伏、创造流畅的通路、易于进入的场所和连续不同的小区意象景观，并塑造强而有力的地标意象，以提高整体景观的自明性及可辨识性，使意象元素的组织符合人类的感知与活动，进而营造独特的意象环境，乃坡地住宅小区计划的重要课题。

总结坡地住宅小区景观意象之塑造应包含环境整体意象和细部单元两种尺度，其设计重点如图05。

坡地住宅小区整体意象之塑造

坡地住宅小区意象之塑造应使居民在进入该区之前，即能接收到提示性的象征讯息和意义。其整体意象因空间组织形态不同而呈现两种风格：低密度的乡村化风格，亦即景观涵容建筑，建筑具"点景"作用，以及高密度的都市化风格，即景观涵容于建筑之中并以建筑为主景。其设计应创造小区整体景观之多样性和趣味性，并保有各种土地使用独特的景观为原则。基本准则如下（图06）：

1. 完善空间组织

住宅小区空间的存在源自使用者与环境间互动的需要，其空间的配置与组织应配合基地既有环境样貌，使之融入现状空间构架，并利用景观美学原理及设计手法，创造引人入胜及具舒适感的氛围，同时，利用景观意象元素创造自明性高且符合人性尺度的空间景观意象。由此，诸如邻里所涉及的区位关系、公共设施、交通动线，以及住宅单

community should focus on the imagery strength by giving with the aim as harmonizing typology and backgrounds. The goal is forming detailed scales, and the target is the landscape elements and the composition to create the harmony and vitality for the community landscape by the contrasting effects made of the differences in forms, materials, colors, proportions and scales from the environment (Fig 04).

In summary, hillside landscape image is actually a three-sided comprehensive phenomenon in which perceptional landscape is influenced by physical landscape and based on imagery landscape. Perceptional landscape creates symbolic images with the mechanism of visual and mental adjustment; furthermore, it constructs the imagery landscape. The physical landscape and perceptional landscape are the foundations of the imagery landscape for hillside residential communities, whereas the forming of imagery landscape and perceptional landscape depends on the presentation of the physical landscape to fulfill the pursuit of functions and visual beauty to create a wonderful residential community environment.

Secondly, the value of the landscape for hillside residential

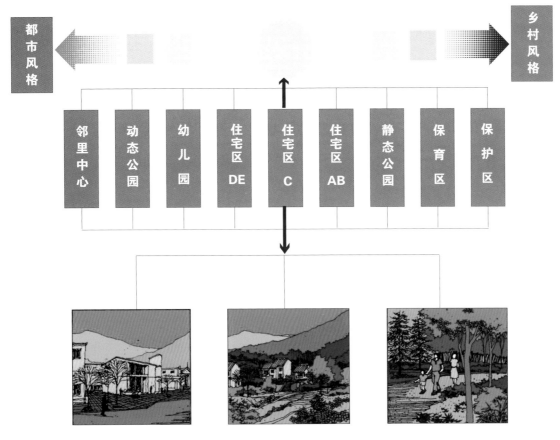

图06 区域风格设计尺度
Fig06 Design scale of local style

元的户外空间配置、尺度、大小及性质等之营造，应使其居民和空间产生互动及认同感。

2. 信道结构分明

信道意象为感知小区系列景观形态变化之主要路径。其沿线序列景观形态以自然元素的多样性和视觉景物的疏密性为主，景致丰富而多变化，人为土地利用形态所造成的意象更具生动性。因此，其组织结构应系统层次分明、清晰易读，并且配合不同功能需要，设计不同的道路形态（图07）。

3. 天空线控制

坡地天空线的剪影因其山势自然起伏和视野辽阔，而常成为住宅小区环境中的视觉焦点之一，相对地容易成为视觉敏感带。因此，应透过容积率、建筑高度限制、植栽配置及建筑外形美化，有效控制人为天际线之发展，并维护与自然天际线协调与和谐（图8）。

4. 小区全面绿化

坡地环境的自然形式，源自地势地貌和水体林木的自然组合，其与住宅小区之人工建筑应通过全面植栽绿化取得和谐性，亦可增加环境生态的稳定性与多样性（图09）。

坡地住宅单元细部意象之塑造

住宅小区整体景观意象在于提供居民认同感与归属感，不仅须藉由各部份空间的共同语汇予以连接，尚须通过相似的外形、材料、色彩及质感等细部元素之变化，强调其特色。其细部设计基本准则包含：

1. 与坡地环境同质的造型

以起伏的山峦为背景，力求建筑造型的调和与美观，并依机能需要采一致性的处理手法。如斜屋顶形式、老虎窗、同比例的开口设计、遮檐收头及阳台处理，以变化突显其多样性，以调和彰显其整体性（图10）。

community is relatively higher for its clear environmental image, less natural linear elements, strong border images, and outstanding physical elements. However, due to the relatively large scale of visual elements and weaker visual continuity and penetration, the major issues for hillside residential community should be: creating fluent circulation by keeping the hillside fluctuation, accessible places and continuous community imagery landscape, forming strong landmark images, and developing the self-identification for the entire landscape to create the unique imagery environment by complying with the imagery elements with mankind's activities and perceptions.

The forming of Landscape Image for hillside residential community should include the two scales for both of entire image and detailed unit, the design principles are as in Fig 05.

The Forming of the entire image for hillside residential community

The forming of the image for hillside residential communities aims to provide the residents the indicating symbols and the corresponding meanings before entering the community area. There are two different styles based on the spatial morphology: Low Density Country Style in which landscape contains architecture, meaning that architectural functions are as decorations; and High Density Urban Style in which the architecture contains landscape, meaning that the architecture is the dominant view. The design should aim to create variety and fun for the community and keep the uniqueness of every place at the same time. The design principles are as follows (Fig 06):

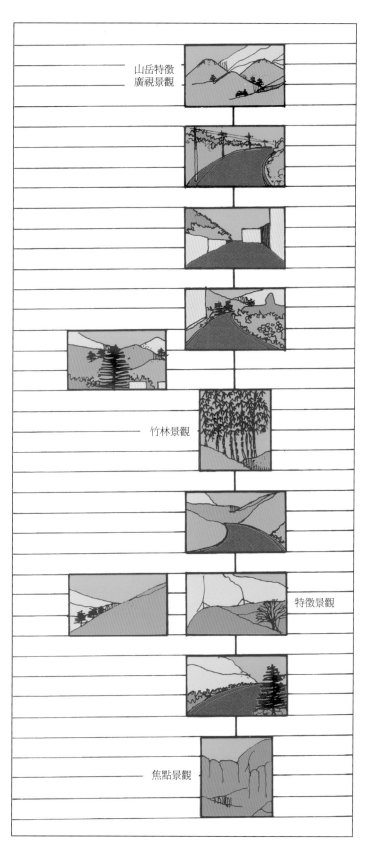

图07 信道意象随着景观型态不同而变化
Fig07 The circulation images change along with the landscape patterns

1. To complete the spatial structure

The residential community comes from the need for the interaction between the residents and the environment whose spatial layout and structure should comply with the existing environmental features of the sites to be integrated into the existing spatial structure. Moreover, it has to apply the landscape design theory and skills to create an attractive and comfortable atmosphere and, in the meantime, to create the spatial landscape image of clear self-identification and humanized scales with landscape image elements. Therefore, neighborhood locations, public facilities, circulations, outdoor layouts, scales, measurements and attributes should all be coordinated to establish the interactions and identifications between residents and space.

2. Clear circulation structure

Circulation is key to the perceptions of the landscape image changes along the community series. The sequential landscape patterns are featured with the variety of natural elements and the density of the visual objects, which is full of rich and various views. Amongst them, the synthetic occupancy patterns are especially with more vitality. Therefore, the structure itself should be clearly systemized, easily interpreted, and designed with different circulation patterns for different functions (Fig 07).

3. Skyline control

The silhouettes of the skylines fluctuating with mountain hills and the grand views as the major visual features for the hillside residential community become the views that the viewers are sensitive about. Therefore, it has to take the regulations of floor area ratio, building heights, vegetation layouts and architectural forms to effectively control the skyline development in order to maintain the harmony with the natural skylines (Fig 08).

4. Overall green coverage for the whole community

The natural form of hillside environment comes from the natural composition of topography, body of water and trees. It should be harmonized with the manufactured buildings inside of the residential community by the application of overall green coverage, so that it furthermore increases the stability and diversity for ecological balance (Fig 09).

The forming of unit detailed image for hillside residential community

The entire landscape image for the residential community aims to provide the residents with the sense of identification and belonging, which takes not only the common vocabulary of the different spaces to connect one another, but also similar detailed elements such as forms, materials, colors and textures to emphasize on the features. The design principles for detailed design are as follows.

1. Similar forms with hillside environment

With the background of fluctuating mountain hills, it should aim to pursue the harmony and beauty of the architectural forms, which are treated with consistency according to the functions, such as tilt roofs, dormers, proportional openings, penthouse headers, and balconies. In other words, it has to present it with variety by differentiation; while to present it with entity by harmonization (Fig 10).

2. Materials and textures reflecting local characteristics

The applications of the building materials should take the natural and local materials in consideration as priority to make the

图 08 坡地住宅小区须谨慎控制人为天际线
Fig08 Effective control of the man-made skyline development to maintain the harmony with natural skylines

图 10 建筑造型力求与坡地环境同质
Fig10 Similar architectural form with hillside environment

图 09 通过小区绿化提高环境生态的多样性
Fig 09 Overall greenery increases the ecological diversity (for the community)

图 11 利用当地材料反映地方特色
Fig11 Materials reflecting local characteristics

2. 反映具有地方特色的材料与质感

建筑和设施材料之使用，应以自然或当地的材料为优先考虑，使之具有地方风格，并配合造型、色彩及其机能形式进行整体设计。材料质感之应用尽量配合现有林木、土壤等粗糙而自然的质感，并以不造成反光或眩光为原则（图 11）。

3. 色彩的巧妙运用

色彩在景观规划中，因受自然环境、植物色彩及其季节性变化，以及光源方向、周期和强度对景观自然元素及人为设施之影响，而有「伪装」及「塑像」两种处理方式美化坡地住宅建筑，使之与周围环境产生和谐感；前者以选择与自然环境色彩较为调和的主调色处理，后者则通过局部鲜明色彩处理大面积的建筑四周墙面、屋顶或凸出的小阳台，如此不仅可达到和谐的整体美感，亦可通过「同中求异」，塑造多样的单元景观。

结论

坡地住宅开发为未来住宅小区发展的主要方向之一，开发得当与否影响景观意象和生态环境甚巨。坡地住宅小区之开发应正视维护坡地自然景观意象之重要性，予以审慎妥善规划，以维护自然景观和生活质量，达致环境生态及自然资源的永续利用。

作者简介：
王小璘／女／博士／教授／朝阳科技大学景观及都市设计系／世界园林杂志总编／台湾
何友锋／男／博士／教授／何友锋建筑师事务所／台湾

Biography:
Xiaolin Wang / Female / Ph.D / Department of Landscape & Urban Design, CYUT / Editor-in-Chief, Worldscape / Taiwan
Youfeng He / Male / Ph.D / He Architectural Studio / Taiwan

entire design full of local characteristics with corresponding forms, colors and functions. The applications of material textures should go with the existing rough and natural textures of trees and earth, and should not be light reflecting in principle(Fig 11).

3. Elaborated applications of colors

Being affected by the natural environments, plants, colors and the corresponding seasonal changes, as well as the light directions, periods and intensities' impacts on natural elements and manufactured facilities, the applications of colors are specified into two ways as "Blending" and "Outstanding" to landscape the buildings on the hillsides. The former way of "Blending" applies the tone, which is blended and harmonized with the colors in the natural environment. Whereas the latter "Outstanding" one will apply the outstanding colors onto the large areas of building walls, roofs or balconies, so that it will not only achieve the harmonized beauty but also form various unit landscapes through the skill of "differentiating from the same".

Conclusion

Hillside residential development has become one of the trends for future residential community development. The appropriateness of the developments will make huge impacts on the landscape images and ecological environments. Therefore, hillside residential community development should emphasize the importance of the preservation of hillside natural landscape images, natural landscapes and living quality with proper evaluations and planning in order to achieve sustainable ecologies and natural resources.

株洲神农城核心区二期景观工程设计
LANDSCAPE DESIGN FOR CORE AREA OF SHENNONG CITY IN ZHUZHOU, STAGE 2

李建伟　　　　　　　　　　　Jianwei Li

项目位置：中国湖南省株洲市　　　　Location: Zhuzhou City, Hunan Province, China
项目面积：67.38hm²　　　　　　　　Area: 67.38hm²
委托单位：神农城建设指挥部部　　　Client: Project Construction Headquarters of Shennong City
设计单位：北京东方艾地景观设计有限公司　Designer: Orient Ideal Design
景观设计：李建伟　　　　　　　　　Landscape Design: Jianwei Li
完成时间：2012年9月　　　　　　　Completion: September, 2012

图 01 神农湖景区一隅
Fig 01 A corner of Shennongcheng Park

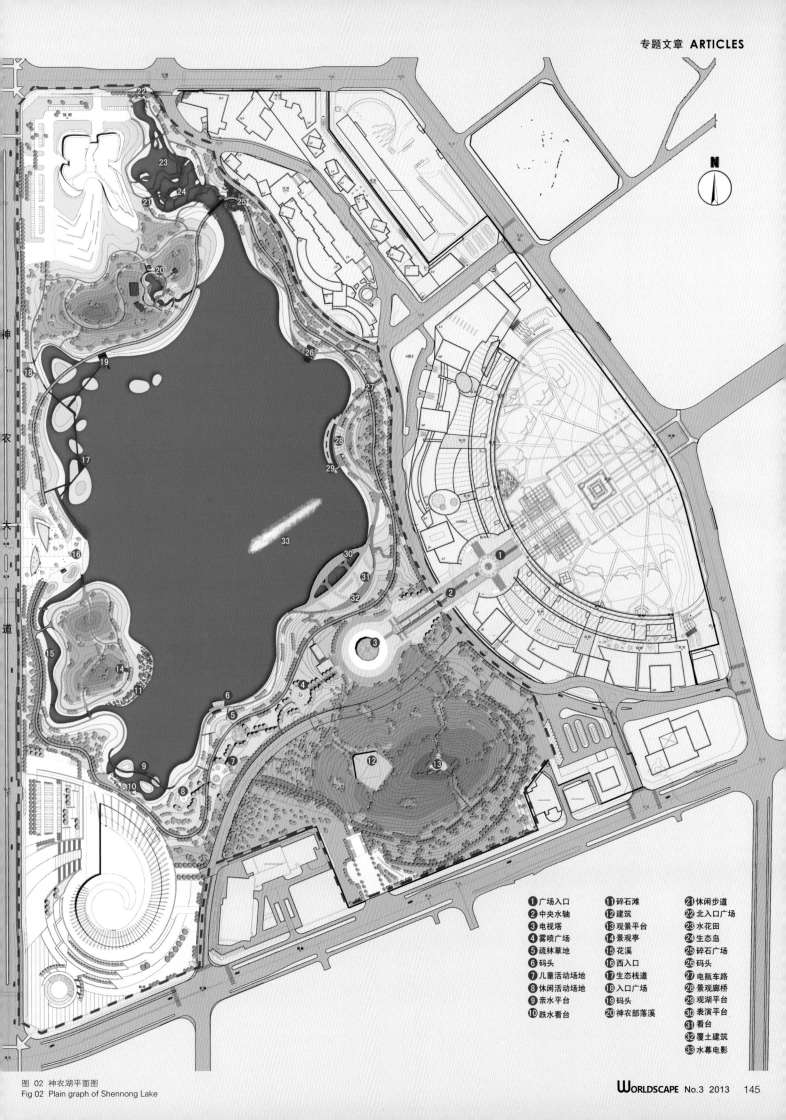

图 02 神农湖平面图
Fig 02 Plain graph of Shennong Lake

① 广场入口　⑪ 碎石滩　㉑ 休闲步道
② 中央水轴　⑫ 建筑　㉒ 北入口广场
③ 电视塔　⑬ 观景平台　㉓ 水花田
④ 雾喷广场　⑭ 景观亭　㉔ 生态岛
⑤ 疏林草地　⑮ 花溪　㉕ 碎石广场
⑥ 码头　⑯ 西入口　㉖ 码头
⑦ 儿童活动场地　⑰ 生态栈道　㉗ 电瓶车路
⑧ 休闲活动场地　⑱ 入口广场　㉘ 景观廊桥
⑨ 亲水平台　⑲ 码头　㉙ 观湖平台
⑩ 跌水看台　⑳ 神农部落溪　㉚ 表演平台
　㉛ 看台
　㉜ 覆土建筑
　㉝ 水幕电影

图 03 神农湖景区花团锦簇
Fig 03 Flowers blooming in the Shennongcheng Park

株洲神农城位于株洲高新技术开发区，项目用地面积67.38hm²，东临炎帝广场，北、西、南大部被村庄和农田环绕（图01-03）。株洲市政府准备将其打造成"神农文化的展示和传播基地，全球华人的炎帝景观中心"。神农城核心区将精心布局与炎帝文化相关的九大建筑与景观，即神农太阳城、神农塔（图04）、神农湖（图05）、神农大剧院、神农艺术宫、神农坛、神农大道、神农渠和神农像。

项目核心区毗邻市级行政中心和湖南工业大学，东接天台路直通河东，西至天元大道，北至珠江北路，南至泰山路。设计以现有炎帝广场和天台公园为核心，以神农文化为主题，在原炎帝广场的基础上，对现有建筑提质改造，对现有城市森林带进行提质升级，并在项目范围内建设生态水系。神农城总体规划为"一带五园"。神农森林带沿神农渠贯穿南北，在这一绿带之上，布置传统文化色彩浓厚的"金、木、水、火、土"五园，"金"为休闲园，"木"为百草园，"水"为湿地园，"火"为炎帝主题园，"土"为农耕园。

Zhuzhou Shennong City is located in the new high-tech development zone of Zhuzhou with a site area of 67.38hm². It is next to Yandi Plaza on the east, while the other three sides are surrounded by villages and agricultural land (Fig.01-03). The Zhuzhou City Council aims to make 'a presentation and dissemination base for Shennong Culture' and 'a Yandi Landscape Centre for All Chinese People'. Major architectural and landscape elements are elaborately arranged in the Core Area of Shennong City, including Shennong Solar City, the Tower of Shennong, Lake Shennong(Fig. 04-05), Shennong Theatre, the Palace of Shennong Art, the Altar of Shennong, Shennong Boulevard, the Canal of Shennong and the Statue of Shennong. (Fig.06-07)

专题文章 ARTICLES

图 04 神农湖水车实景图
Fig 04 Real photo of the water wagon of Shennong Lake

图 05 神农湖实景图
Fig 05 Real photo of Shennong Lake

图 06 设计效果图
Fig 06 Rendering Picture

图 07 设计效果图
Fig 07 Endering Picture

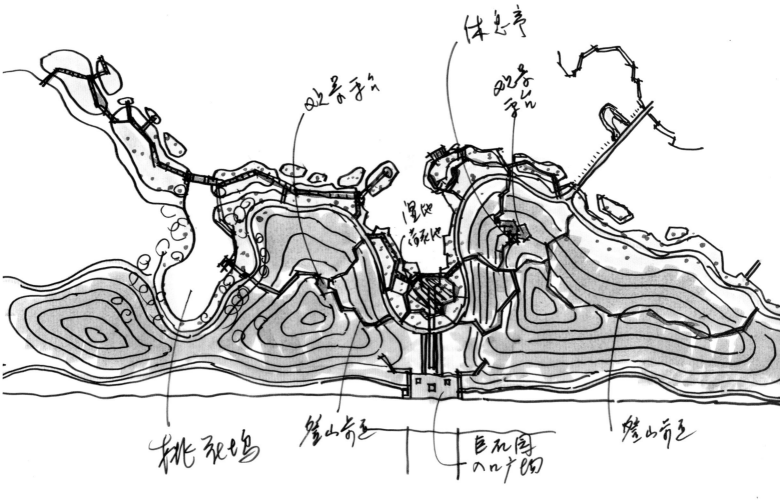

图 08 神农山手绘图
Fig 08 Sketch of Shennongcheng Park

方案整体设计思想是以"神农文化"为核心理念，展现株洲水映山灵、锦绣绿城的城市新形象。依据株洲建设"两型"的总体思想，根据基地自然条件，结合周边地块功能性质，划分景观功能空间，形成集文化、旅游、商业于一体的新型城市滨水开放空间，提升株洲城市的名片形象（图06-07）。

株洲神农城景观方案根据基地自然条件，结合周边地块功能性质进行功能分区，形成集文化、旅游、商业于一体的新型城市滨水开放空间。方案设计大气磅礴，施工也是本着最大限度体现设计思想的目标，注重细节、精益求精，把设计方案落实到位，保证了实施效果（图08-09）。

株洲神农城为突出未来之城的特质，从设计到施工，神农城都将贯穿低碳理念。用设计者的话说："神农城每一个细节，小到一块石头，大到水循环，都要体现低碳理念。"为此，城内大量采用太阳能、地热等新能源、新技术；建筑结构及建材构建上多采用隔热、保温等国家倡导的技术和材料；在水资源利用上，引入污水处理尾水和中水回收循环使用技术，实现新能源、新技术与城市建设的良好融合。

株洲神农城是一个具有得天独厚的地理区位优势与环境景观优势的项目，该项目通过重塑原炎帝广场与原规划中的天台公园地块一并纳入神农城项目建设规划红线范围内的功能与形象，不仅仅是一次城市提质改造、更新的项目，建成后的神农城将是现代都市生活的休闲福地和城市中心区极具诱惑力的生态走廊，它将引领正处于经济结构转型时期的株洲城市新的生活方式。□

The core Area of this project is located nearby the municipal administration centre and the Industrial University of Hunan. The site is intersected by Tiantai road on the east, Tianyuan Boulevard on the west, Zhujiang North Rd on the north, and Taishan Road on the south. The project was designed with the existing Yandi Plaza and Tiantai Park as a core. Based on the existing plaza, the quality of the existing architecture was upgraded, the existing urban forest was enhanced, and an ecological hydrologic system was established. The concept of 'one belt and five gardens' is the basis for the planning within Shennong City. Shennong Forest belt runs along the Canal of Shennong from north to south. Five gardens or parks are arranged along this green belt with themes according to the Chinese Five Elements, including Gold, Wood, Water, Fire and Earth. The five theme parks respect local knowledge and endow the traditional cultural elements with functional uses. 'Gold' is a recreational park, 'Wood' is a herbal garden, 'Water' is a wetland park, 'Fire' is a historical garden of Yandi culture, and 'Earth' is a community garden.

The goal of this project is to focus on the culture of Shennong in order to present a new urban image of Zhuzhou, where clear water mirrors misty mountains, and the city is refreshed by beautiful green space. According to the policy of "resource

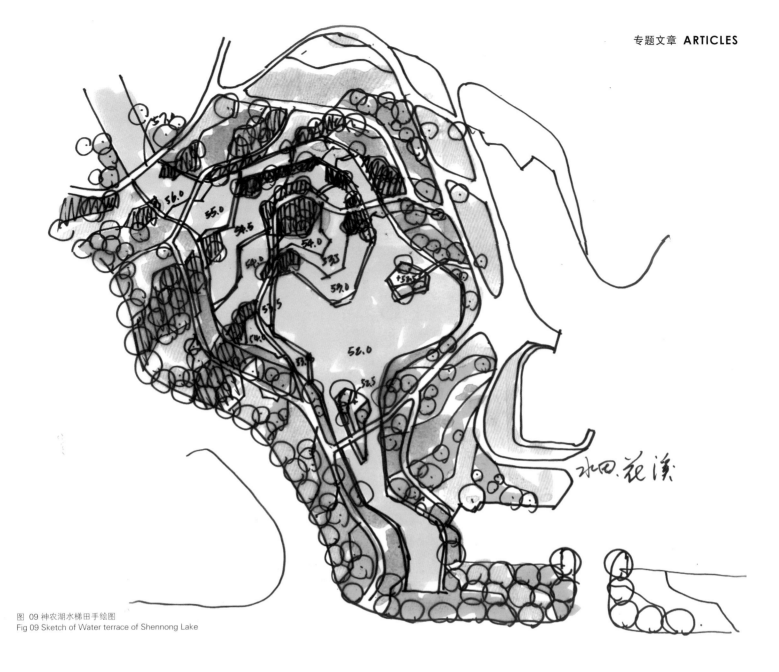

Fig 09 Sketch of Water terrace of Shennong Lake

conservation and an environmentally-friendly society", the design follows the local existing conditions, including the function, context and character of the various urban zones. In order to enhance the iconic identity of Zhuzhou, areas of the landscape have been rearranged to establish a new urban waterfront open space, with the goal of promoting cultural conservation, tourism and business.

The landscape design of Shennongcheng Park, in Zhuzhou city follows the existing natural conditions and conforms to the existing functions of each plot by integrating with the functional characteristics of the surrounding land. The design presents a new model of urban waterfront open space in its particular integration of cultural, tourist and commercial functions. The landscape design is majestic and spectacular, while the construction is precise and elaborate. (Fig.08-09).

'City of the future' is the main theme of this project. From design to construction, Shennongcheng maintains the principle of low carbon design. 'Whether it be a tiny stone or a drop of recycled water, every detail should reflect the low carbon principle' the designer said. Therefore, the city made extensive use of solar, geothermal and other new energy sources and new technologies, including building structures and constructions with more effective insulating techniques and materials. In terms of water use, this project makes use of sewage disposal and intermediate water recycling technologies to achieve completed integration between new resources, new technologies and urban construction.

The Shennongcheng site provides natural geographical benefits, which the design takes full advantage of. By regenerating the landscape of Yandi plaza and Tiantai Park, a leisure centre is provided for local residents and an ecological corridor is created in the city center. This urban landscape renewal project will lead people's new life style when it is built.∎

作者简介：

李建伟 / 男 / 风景园林师 / EDSA Orient 总裁兼首席设计师 / 中国

Biography:

Jianwei Li / Male / Landscape Architect / President and Chief Designer of EDSA Orient / China

查尔斯·沙（校订）

English reviewed by Charles Sands

源树景观（R-land）是国内顶级的专业 环境设计机构。自2004年成立以来，通过不懈的努力，在景观规划、公共空间、旅游度假、主题设计等领域都获得了傲人的成绩，特别是在高端地产景观的咨询及设计方面，处于绝对的领先地位。

源树景观（R-land）的设计团队中汇集了大量的国内外景观设计精英，其主要设计人员都曾在国内外高水平设计单位中担任重要职务，严格的设计流程确保了每一项设计作品的完美呈现。

源树景观（R-land）历经数年，已完成了数百项设计任务，其中：河北省邯郸市赵王城遗址公园、中关村创新园、山东荣成国家湿地、西安大唐不夜城、北京汽车博物馆、龙湖"滟澜山"、天津团泊湖庭院、招商嘉铭珑原、远洋傲北、中建红杉溪谷、西山壹号院等若干项目均已建成并得到各界的广泛认可。

源树景观致力于最高品质的景观营造，力求为合作方提供最高水准的设计保障。

Add: 北京市 朝阳区朝外大街怡景园 5-9B（100020） Tel:（86）10-85626992/3　85625520/30　Fax:（86）10-85626992/3　85625520/30 - 5555　Http://www.r-land.com

R-land

Beijing -Tianjin -Tokyo -Sydney　　YS Landscape Design

http://www.r-land.cn　源树景观

景观规划 Landscape Planning　　公共空间 Public Space　　居住环境 Living environment　　主题设计 Theme Design

竞赛作品 / WORKS FROMS

渗透现象
——呼和浩特市南湖公园湿地科普景区（方案一）
THE WATERFRONT LANDSCAPE DESIGN OF THE HOHHOT SOUTH LAKE PARK(PLAN I)

中文标题：渗透现象——呼和浩特市南湖公园湿地科普景区（方案一）
组　　别：本科
作　　者：刘昶
指导教师：许大为
学　　校：东北林业大学
学科专业：风景园林
研究方向：风景园林规划设计
分　　类：2012"园冶杯"风景园林（毕业作品、论文）国际竞赛设计类作品三等奖

Title: the waterfront landscape design of the Hohhot south lake park（plan I）
Degree: Bachelor
Author: Chang Liu
Instructor: Dawei Xu
University: Northeast Forestry University
Specialized Subject: Landscape Architecture
Research: Landscape Architecture Planning and Design
Category: The Third Prize in Design Group of 2012"Yuan Ye Award"International Architecture Graduation Project/Thesis Competition

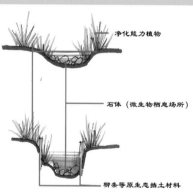
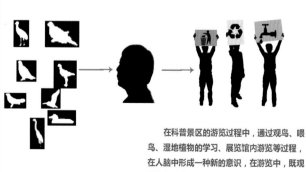

图 01 主题思想图
Fig 01 Concept Design

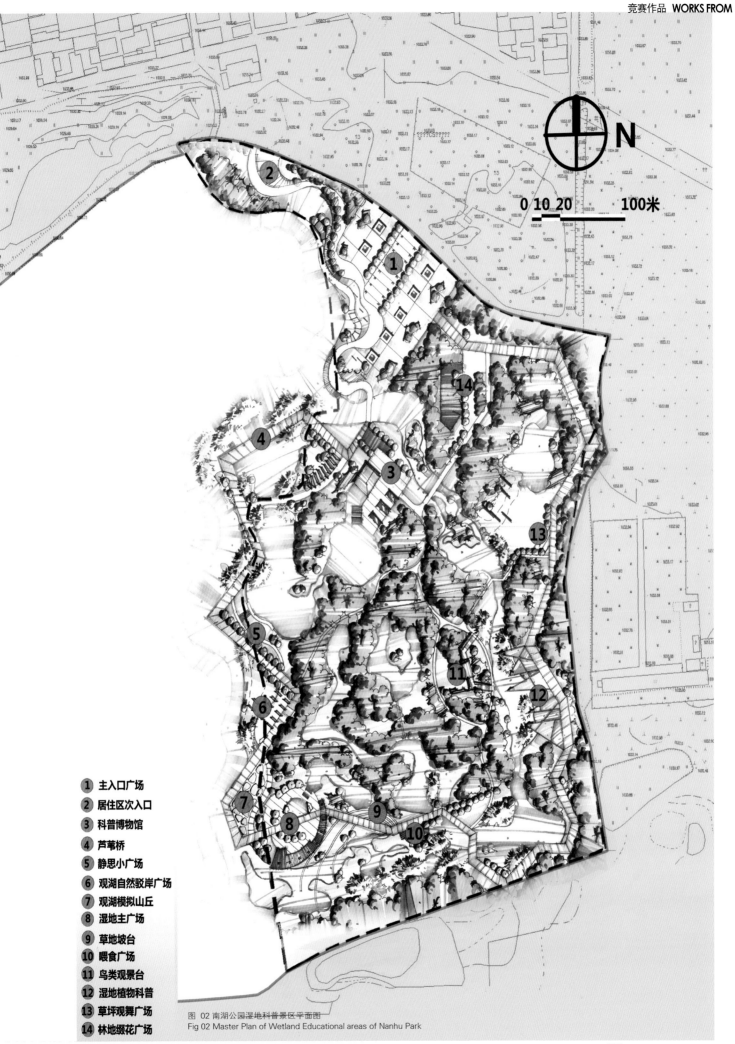

图 02 南湖公园湿地科普景区平面图
Fig 02 Master Plan of Wetland Educational areas of Nanhu Park

1. 主入口广场
2. 居住区次入口
3. 科普博物馆
4. 芦苇桥
5. 静思小广场
6. 观湖自然驳岸广场
7. 观湖模拟山丘
8. 湿地主广场
9. 草地坡台
10. 喂食广场
11. 鸟类观景台
12. 湿地植物科普
13. 草坪观舞广场
14. 林地缀花广场

索引图

① 流水　⑥ 图腾柱
② 浅水水池　⑦ 湿地景观
③ 树阵座椅　⑧ 坡度步道
④ 特色景观树　⑨ 碎石小道
⑤ 木栈　⑩ 坡道座椅广场

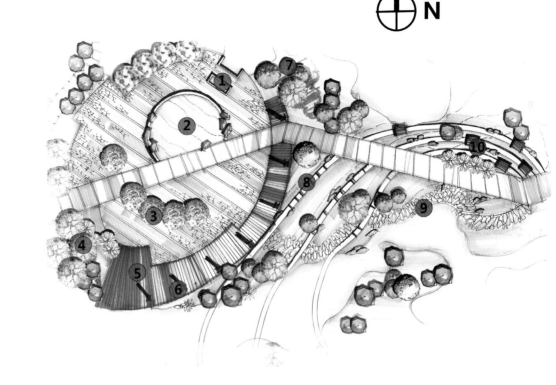

图 03 科普广场局部平面图
Fig 03 Plan of Science Plaza

作品简介：

本设计结合内蒙古自治区呼和浩特市南湖湿地公园特色，以"渗透"为主题思想（图01），通过规划滨水陆地和岛屿两部分，创建一个湿地科普景区（图02），反映湿地科普教育主题。充分考虑环境空间尺度、功能和交通组织等的基础上设置了科普广场（图03-04）、科普博物馆、观鸟喂鸟广场和湿地植物驳岸（图05）等空间为人群创造一个湿地科普学习环境。设计时从整体景观考虑，同时体现区域环境特征（图06-07）。在具体景观及空间的创造上，符合时代要求和现代人的审美情趣，满足人们休闲、娱乐、科普和赏景的需求（图08）。□

Instruction:

The plan is based on the characteristics of NO.9 Station Park in Huhehaote, Inner Mongolia Autonomous Region and is based on the concept of "permeate"(Fig.01). It creates an educational wetland scenic spot(Fig.02) by planning waterside areas and islands to reflect the theme of scientific education. The design fully considers the spatial scale, function and transportation requirements of the site. Features include a science plaza (Fig.03-04) and museum, and a bird square with wetland shore plants(Fig.05) to provide people an educational environment. The design also integrates with the surrounding landscape and the broader regional landscape (Fig.06-07). Through modern aesthetics, the design provides a place for leisure, amusement, education and sightseeing (Fig.08).■

查尔斯·沙（校订）
English reviewed by Charles Sands

图 04 科普广场局部剖面图
Fig 04 Section of Science Plaza

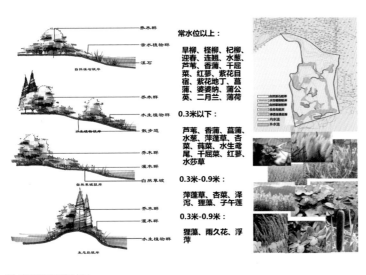

图 05 驳岸规划设计图
Fig 05 Plan of the Revetment design

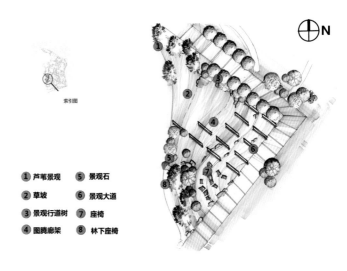

图 06 柳岸余荫局部平面图
Fig 06 Plan of Willow Shadow

图 07 柳岸余荫局部剖面图
Fig 07 Section of Willow Shadow

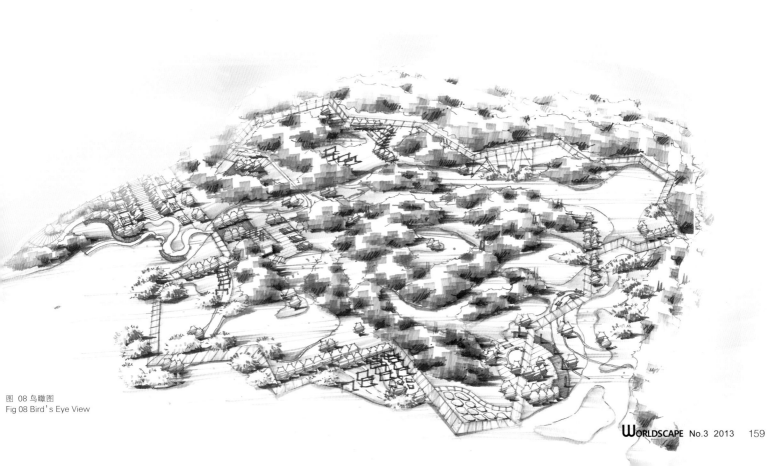

图 08 鸟瞰图
Fig 08 Bird's Eye View

滋育
东北农业大学校园广场景观规划设计
MOISTEN AND NOURISH
NORTHEAST AGRICULTURAL UNIVERSITY CAMPUS SQUARE LANDSCAPE PLANNING AND DESIGN

中文标题：滋育——东北农业大学校园广场景观规划设计	Title: Moisten and nourish——Northeast Agricultural University campus square landscape planning and design

中文标题：滋育——东北农业大学校园广场景观规划设计
组　别：本科
作　者：万亿
指导教师：胡海辉 曲畅
学　校：东北农业大学园艺学院
学科专业：风景园林
研究方向：景观规划设计
分　类：2012"园冶杯"风景园林（毕业作品、论文）
　　　　国际竞赛设计类作品三等奖

Title: Moisten and nourish——Northeast Agricultural University campus square landscape planning and design
Degree: Bachelor
Author: Yi Wan
Instructor: Haihui Hu Chang Qu
University: Northeast Forestry University
Specialized Subject: Landscape Architecture
Research: Landscape Planning & Design
Category: The Third Prize in Design Group of 2012"Yuan Ye Award"International Architecture Graduation Project / Thesis Competition

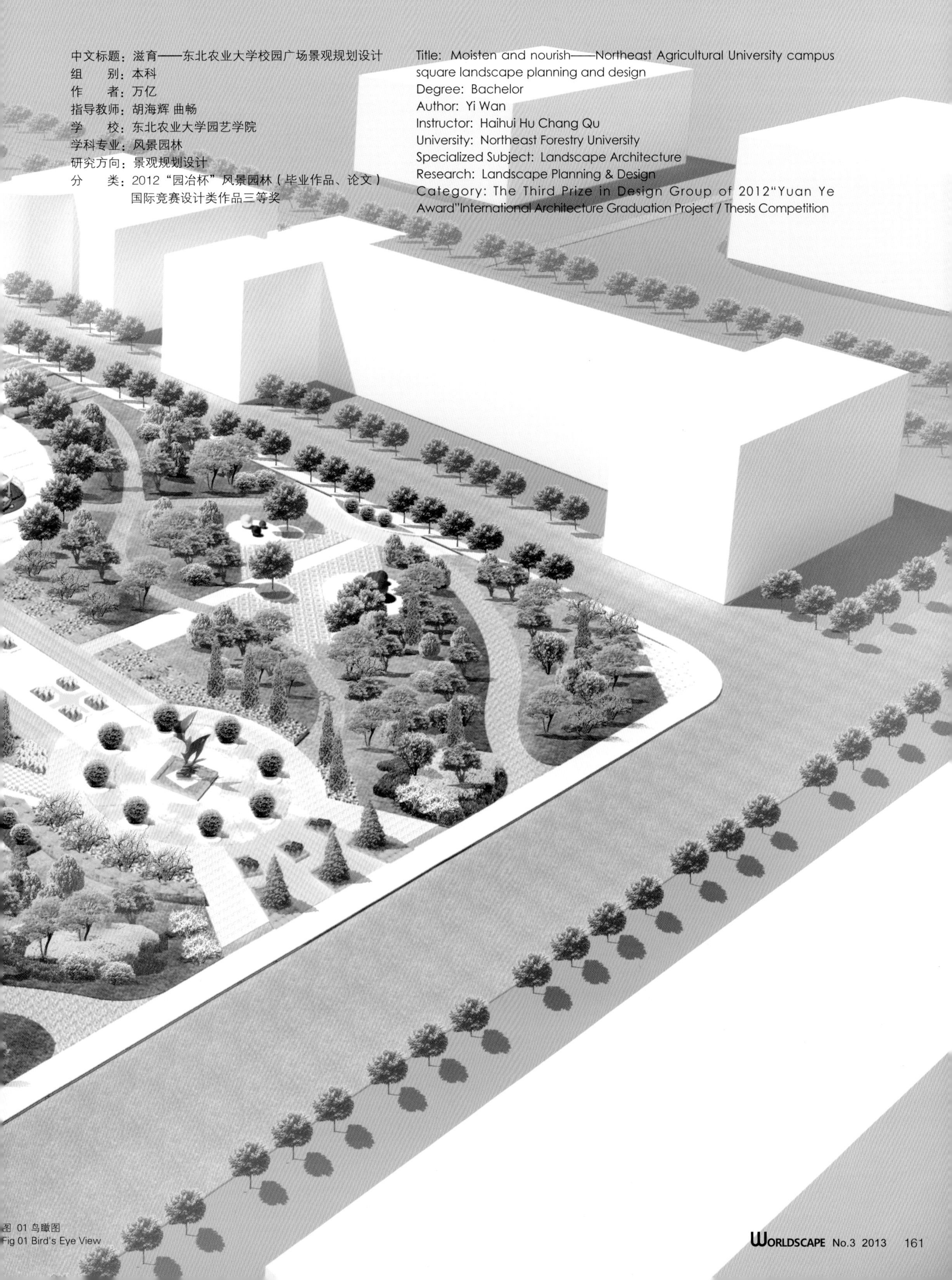

图 01 鸟瞰图
Fig 01 Bird's Eye View

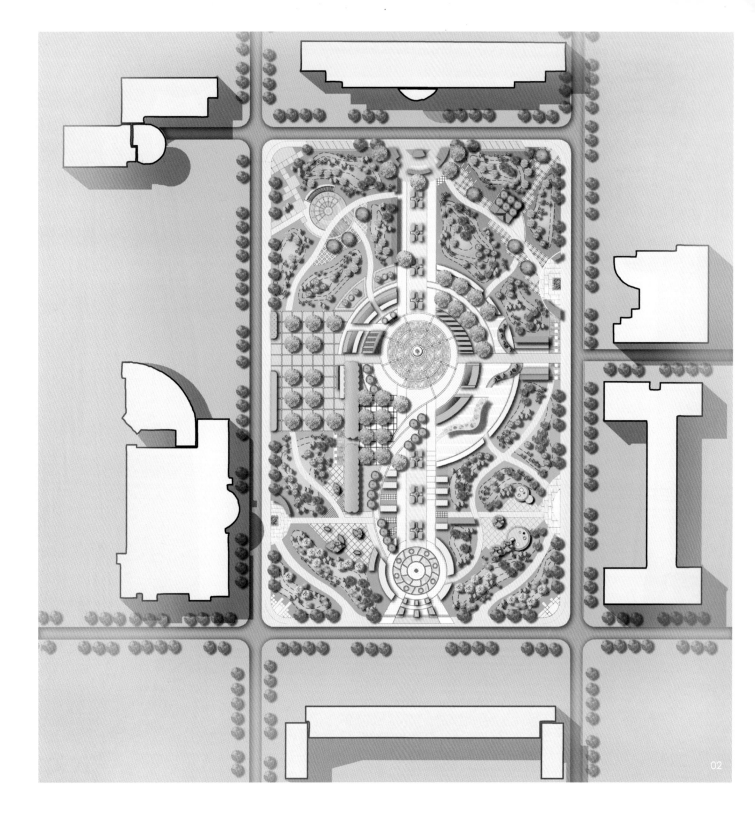

作品简介：

本设计以"滋育"为主题，即滋润、养育，孕育生命与希望。既代表了人的成长过程（孕育与教育），也体现了植物的生长周期，即从种子在土地中萌发，到植物的叶、根、茎、花朵和果实生长的过程（图01-02）。方案中的各景点分别以春花、夏田、秋叶、冬果为一种意象，通过植物景观设计，营造出缤纷多彩的季相景观（图03-07）。此外，"滋育"主题也体现在将地形高差与雨水收集技术相结合。这一方案融入鲜明的农科高校特色文化，为学生提供舒适宜人的游憩与交流活动空间。

Introduction:

With the theme of "moisten and nourish", this landscape design symbolizes life and hope (Fig.01). It stands for the growth process of people (civilization and education), but also signifies the growth cycle of plants, from seed germination in the soil, to leaves, roots, stems, flowers and fruit (Fig.02). The project employs spring flowers, summer fields, autumn leaves and winter fruit as representative images to create a variety of seasonal plant landscapes (Fig.03-07). In addition, the design employs elevation changes in the terrain to harvest rainwater. This project reflects the distinctive characteristics and culture of the agricultural university to provide students with an area for pleasant recreation and comfortable social interaction.

查尔斯·沙（校订）
English reviewed by Charles Sands

图 02 总平面图
Fig 02 Master Plan

图 03 "莺飞蝶舞"区效果图
Fig 03 Rendering of "Spring"

图 04 "粉妆玉砌"区效果图
Fig 04 Rendering of "Winter"

图 05 光阴中心广场效果图
Fig 05 Rendering of "Time"

图 06 "金风玉露"区效果图
Fig 06 Rendering of "Autumn"

图 07 "一碧万顷"区效果图
Fig 07 Rendering of "Summer"

哈尔滨市九站公园滨水区环境设计
THE WATERFRONT LANDSCAPE DESIGN OF THE NO.9 STATIONS PARK

中文标题：哈尔滨市九站公园滨水区环境设计
组　　别：本科
作　　者：王婧
指导教师：毕迎春
学　　校：东北林业大学
学科专业：环境艺术设计
研究方向：景观规划、景观设计
分　　类：2012"园冶杯"风景园林（毕业作品、论文）国际竞赛设计类作品三等奖

Title: The Waterfront Landscape Design of the NO.9 Stations Park
Degree: Bachelor
Author: Jing Wang
Instructor: Yingchun Bi
University: Northeast Forestry University
Specialized Subject: Environmental Art Design
Research: Landscape Planning & Design
Category: The Third Prize in Design Group of 2012"Yuan Ye Award"International Architecture Graduation Project/Thesis Competition

图 01 九站公园全景图
Fig 01 Sketch-up model of NO.9 Stations Park

图 02 总平面图
Fig 02 Master Plan

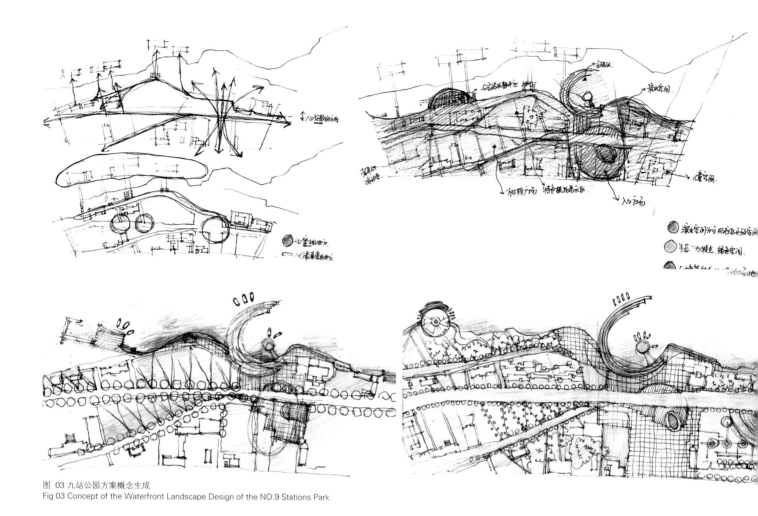

图 03 九站公园方案概念生成
Fig 03 Concept of the Waterfront Landscape Design of the NO.9 Stations Park

图 04 九站公园局部效果图
Fig 04 Sketch of NO.9 Stations Park

05 九站公园局部效果图
Fig 05 Sketch of NO.9 Stations Park

作品简介：

本项目试图从阅读场地，感受自然过程，体会场地的精神出发，通过景观的形式阐述了对传统与现代，人与自然关系的理解。以滨水区规划理念为依据，再现哈尔滨滨水区历史文化（图01-04），对开放空间进行精心设计，使游人共享滨水的乐趣和魅力（图05），感受哈尔滨历史文化的变迁和历史文明的发展。希望通过该设计创造宜人滨水环境（图06-08）。对改善城市日益恶化的水域生态环境，恢复水文化生机做一次有益的尝试，实现真正意义上的生态发展和可持续发展。

Introduction:

We begin by going over the space, feeling the natural processes, understanding the spirit of the place, figure out the relationship between the traditional and the modern, and man and nature by means of the landscape. According to the idea of public waterfront planning, we reveal the historical culture of the waterfront of Harbin (Fig.01-04). By designing an elaborate open space, people are able to enjoy the delight and charm of the waterfront space, feeling the historic cultural transition and the expansion of civilization through history (Fig.05). We propose improving the deteriorating waterfront environment (Fig.06-08), with the goal of recovering the vitality of the waterfront, and achieving ecological and sustainable development.

图 06 鸟瞰图
Fig 06 Bird's Eye View

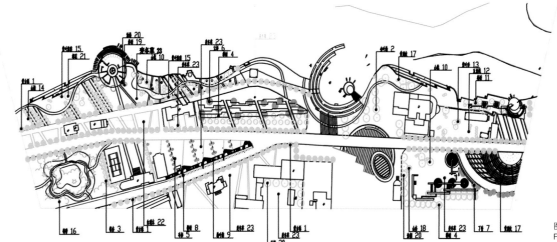

图 07 植物配置图
Fig 07 Plant Configuration Diagram

竞赛作品 WORKS FROM

图 08 节点详图
Fig 08 Node Details

查尔斯·沙（校订）
English reviewed by Charles Sands

垂直绿化工法
VERTICAL GARDENING CONSTRUCTION TECHNIQUES

谢宗钦　　　　　　　　　Tzung-Chin Hsieh

摘要

因应近百年科技不断的进步，人类大量的排放水蒸气、二氧化碳、臭氧、甲烷、氮氧化物等温室气体，致使大气的温室效应增强，进而使地球的温度也不断的上升。

为降低温室效应的不断增强，全球都在致力于如何减少各类温室气体的排放，但若要经济持续发展，要达成节能减碳的目标，在除个人方面于日常生活中，要养成随手关灯、关水的习惯，尽量多使用大众交通工具以减少能源的使用量外。能源的开发方面，亦要多发展如太阳能、地热、风力、水力、潮汐及氢燃料等新能源以减少二氧化碳气体的产生。

解决地球温室效应最根本且最有效的办法是停止砍伐天然林，大量的增加都市绿化面积，然而，当今在都市发展快速之下，都市的土地寸土寸金，都市中大楼到处林立，可用于绿化的面积相对减少，且大楼表面的建材更为大气带来更严重的热岛效应，为有效的增加绿化面积，在有限的平面空间内，将绿化由平面转变成垂直实为最符合成本效益及速效性的方式。

传统的垂直绿化是在平面栽植攀爬植物，让植物慢慢的生长，如此至少需要数年之久才能达到有效的绿化效果，新式的工法是以不织布或模块系统像堆积木的方式堆栈起来，植栽直接生长于不织布或模块槽盒之中，即可立即达到垂直绿化的效果。

垂直绿化的技术，除如何选用合适的植物种类外，系统中为能让所有的植栽都能平均得到水的供给，一般的设计都讲究在如何使水能平均的供给，却忽略了如何使水安全排放，结果因为地心引力的关系，上层的排放水逐层的流经底下的植栽区，如此不但容易让水沉积在下层的植栽区，而造成上下供水不均，亦容易造成土壤病害的交互感染。

这样就如同居住在大楼20楼的住户将使用过的水流给19楼的住户使用，然后逐层的流到底层再排出，如此住户中就很容易造成传染病的交互感染。

以下所介绍之垂直绿化系统，在设计时除考虑到组装时的快速性及日后维护上的方便性外，更以人的居住环境角度去思考植物的生存环境，垂直绿化中的所有植物其所需要的水就如大楼里的人，都取自于顶楼的干净水，每一株植物灌溉后的排放水也如同建筑物有专属的排水管直接排出，如此不但不会使水沉积在下层，造成上下供水不均的问题，彼此间更不会产生土壤病害的交互感染。

Introduction

Large amounts of greenhouse gases (i.e. water vapor, CO_2, CH_4, N_2O, etc.) have been produced by humans through rapid industrial growth over hundreds of years, and this has eventually resulted in increasing ambient temperatures. In order to sustainably develop our environment, it is necessary to relieve the greenhouse effect. For this purpose, each country is devoted to reducing the production of greenhouse gases. In addition to our daily energy saving, using alternative energy and increasing vegetation area can efficiently reduce the production of greenhouse gases.

However, it is difficult to increase vegetation in cities nowadays because land is too expensive to purchase. The surface materials of high buildings generate an urban heat island effect and further increase the ambient temperature. In order to efficiently increase vegetation area within the limited space available in cities, converting gardening from horizontal planes into vertical gardening is one of the fastest and most economical ways.

Traditional vertical gardening, which grows vines along a wall or lattice in a flat plane, is time-consuming. New vertical gardening construction techniques directly grow plants in mats or module containers and then assemble them like bricks to make vertical garden (green wall) instantly.

After plant choice, the most important aspect of designing vertical gardening systems is creating an efficient water supply. Most designs emphasize equal water supply for each plant in the system, but they ignore appropriate water drainage. Here, water will drain from the upper layers of the vertical system to the layers below them, layer by layer. As a result, the water supply for each plant in the vertical gardening system will become unequal: plants in lower layers will get more water than those in upper layers. Moreover, this kind of water drainage design will make soil-borne plant diseases transmit from infected plants to healthy plants easily.

The following introduced vertical gardening system is designed to solve these issues. In addition to its fast assembly and easy maintenance, our vertical gardening system has the advantage of being designed from the perspective of plants themselves. Plants in our vertical gardening system are treated like humans living in a high building: they each get an equal share of clean water from a water tank located on top of the building, and have their own separated water drainage pipes which directly flow into a water collection basin in the bottom. Thus, our vertical gardening system can eliminate inequalities in water supply between upper layers and lower layers and reduce the transmission of soil-borne diseases between plants.

新材料 NEW MATERIALS

一、产品名称：三孔立体植栽槽

二、功能、特性、材料

(1) 每一植栽槽单元内设有灌溉管道及排水管道，安装快速便捷，各植栽槽之灌溉、排水无须另行配管。

(2) 每一植栽槽单元各具有灌溉管道与排水管道，灌溉与排水有分流的功能，土壤病害不会交互感染。

(3) 独立的灌溉系统，确保上下各植栽槽能平均供水。

(4) 隐藏式的灌溉、排水管道，不易受到外力破坏。

(5) 独立的植栽灌溉系统，植栽养护更换容易。

(6) 材质：耐候性pp材质。

(7) 示意图（图01）

1. Product name: Three-hole vertical planting container

2. Features:

(1) Each unit of the container has built-in water input and drainage systems. Moreover, it is easy to assemble and does not need additional pipe installation.

(2) Each unit of the container has independent built-in water input and drainage systems, which will prevent the spread of soil-borne diseases.

(3) The independent built-in water input and drainage systems also ensure that each planting container receives an equal amount of water.

(4) Changing individual plants is easy because of the independent built-in water input and drainage system for each planting container.

(5) The hidden design of built-in water input and drainage systems reduces the potential for damage by external forces.

(6) Material: Weather-resistant polypropylene

(7) Diagram (Fig 01)

图 01 三孔立体植栽槽示意图
Fig 01 Illustration of the three-hole vertical planting container

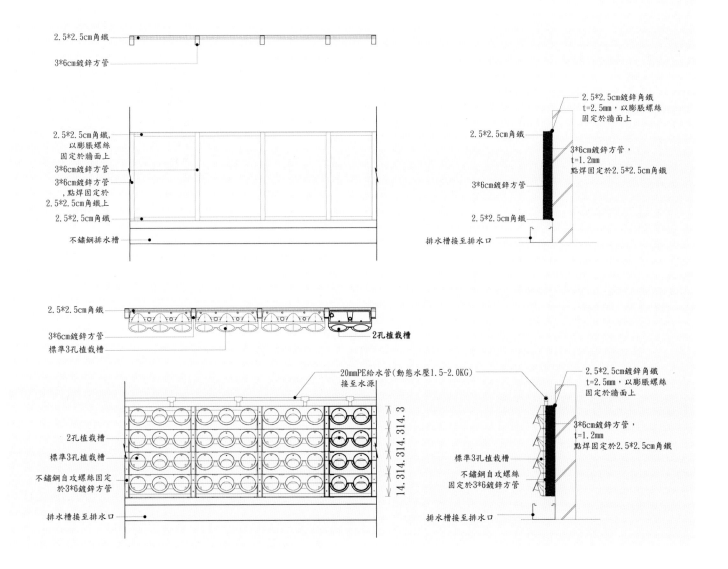

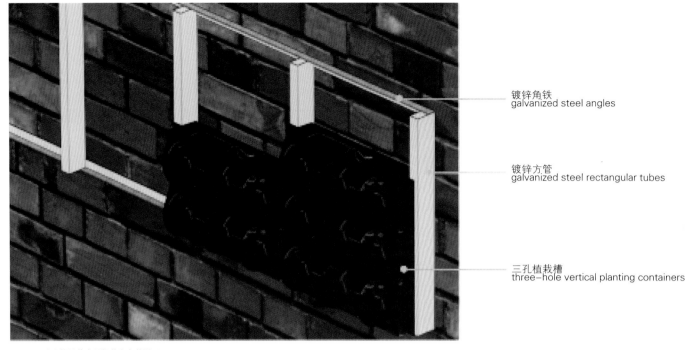

图 02 三孔立体植栽容器组装和安装说明
Fig 02 Instruction of assembly and installation of three-hole vertical planting containers

三、施工工法

3.1 施工说明

(1) 以镀锌角铁及镀锌方管做为植生墙的外围架构来固定植栽槽（图02）。

(2) 植栽槽的安装，可以暂时以四层为单位先锁一螺丝做为假固定，待完成试水确定无误后再全部锁上螺丝。

(3) 按装自动供水系统、控制阀及定时器。

(4) 植栽放置前须加装植栽网套，再依配置图所示放入植栽槽体，并装设挡板防止强风吹落。

3.2 系统维护（排水、进水、槽体）

(1) 做放水测试，确认供水管线的 PE 管接合处，是否有局部漏水，若有漏水需局部更换材料，以确保供水不外泄，及供水管的水压维持。

(2) 检视每个三孔植栽槽的滴孔是否有正常供水，如无漏水，供水管水压正常时，滴孔仍无正常出水，检视滴孔是否有堵塞、分流滴管是否松脱，可以用手逐一将分流滴管再次套紧。如三孔植栽槽在确认为滴管套组堵塞，则须更换一组滴灌管套组。

(3) 控制器宜选用电子式定时器，定期检查时要检视定时器是否能依排程运作，定时供水时间是否达到所需时间。

(4) 每一排花箱底部设有清洗阀，用以排放管中的杂物，如该清洗阀在正常运作下会漏水，就是内有杂物，应拆开检查。

(5) 水源过滤器需定期拆开清洗，视水质而定，多为 2~3 个月清洗一次。

3. Construction and maintenance

3.1 Construction steps:

(1) Galvanized steel angles and galvanized steel rectangular tubes are used to make a vertical garden system framework for fixing planting containers (Fig 02).

(2) To assemble vertical planting containers, water input and drainage should be tested and containers should be loosely attached to each other at first, using one screw to fix each container for every four layers of planting containers. If the water input and drainage systems work well, the rest of the screws can be fastened to fix the containers into the framework.

(3) Next, the automatic water supply system, the control valve, and the timer are installed.

(4) Plants are placed into net pots and inserted into three-hole vertical planting containers according to the layout drawing. Finally, block boards are installed on the containers to prevent plants from blowing down.

3.2 Maintenance of the vertical gardening system:

(1) The water supply is turned on and the junction of the water supply PE pipe at the top end of the system is tested for evidence of leakage. If there is leakage, leaky junctions need to be replaced to avoid leakage and maintain a stable pressure in the water supply pipe.

(2) The water drip hole of each three-hole vertical planting container should be observed to confirm that the water supply flows through normally. If there is no leakage but no water comes out from the water drip hole, the water drip hole may be clogged or the water drip pipe may be loose. It is possible to manually check water drip pipes and wedge them to the water drip emitters tightly. The drip irrigation pipe set of the three-hole vertical planting container can be replaced if it is clogged.

(3) It is best to use an electronic control for regulating the water supply and to check it periodically to confirm that it works as schedules and reaches the time set up for water supply.

(4) There is a clean valve at the bottom of each column of the vertical gardening system. If it leaks under normal condition, this suggests it is clogged. In this situation, it is advisable to open the bottom of the vertical gardening system and clean the clean valve in it.

(5) The water filter of the water supply source should be cleaned periodically. In most cases, cleaning should be scheduled every two to three months.

台北市社会教育馆
TAIPEI CULTURE CENTER, TAIPEI, TAIWAN

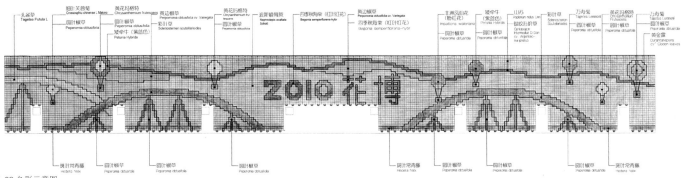

图 03 色彩示意图
Fig 03 Layout drawing

位于台北市八德路上的社会教育馆，为大家熟悉的艺文空间，为响应「2010 年台北国际花卉博览会」，于相关动在线，作城市绿美化（图 03-05）。

名　　称：台北市社会教育馆绿墙
位　　置：台北市信义区八德路三段 25 号
空间型态：公用空间
绿化面积：约 192m²
设置目的：长久性墙面绿化
使用植栽：波士顿蕨、彩叶草、斑叶长春藤、四季秋海棠等

The Taipei Culture Center is located on Bade Rd in Taipei City and is a venue for art displays and cultural activities for the people of Taipei. In support of the 2010 Taipei International Flora Exposition, the Taipei Culture Center installed vertical gardening systems along surrounding paths (Fig 03-05).
Name: Taipei Culture Center Green Walls

Location: No.25, Sec. 3, Bade Rd., Xinyi Dist., Taipei City 106, Taiwan (R.O.C.)
Space type: Public
Area size: ≈ 192m²
Purpose: Long-term vertical gardening
Plants: Nephralepis exaltata cu. Bastaniensis, Solenostemon scutellarioides, Hedera helix, Begonia semperflorens-cultorum, etc.

图 04 台北市社会教育馆完工照片
Fig 04 The actual green walls based on the layout drawing in the Taipei Culture Center surrounding

图 05 台北市社会教育馆后续植栽配置
Fig 05 The green walls with different design by changing plants in the Taipei Culture Center Surrounding

京站时尚广场
QSQUARE, TAIPEI, TAIWAN

位于台北交通的枢纽的京站时尚广场，为近几年来热门的时尚百货公司。入口处，一片绿意盎然的绿墙，使人彷佛进入世外桃源般（图 06-08）。

名　　称：京站时尚广场绿墙
位　　置：台北市大同区承德路一段1号
空间型态：商用空间
绿化面积：约400m²
设置目的：长久性墙面绿化使用植栽：文竹、凤尾蕨、鸭趾草、合果芋、山苏、蔓绿绒、黄金葛、千年木等

Q Square, located in the center of Taipei traffic, is one of the most popular department stores in Taipei. The entrance is decorated by green walls with flourishing plants (Fig 06-08).
Name: Qsquare Green Walls
Location: No.1, Sec. 1, Chengde Rd., Datong Dist., Taipei City 103, Taiwan (R.O.C.)
Space type: Business
Area size: ≈ 400m²
Purpose: Long-term vertical gardening

图 06 西侧挑空区绿墙 1
Fig 06 Green walls in west side of Qsquare 1
图 07 西侧挑空区绿墙 2
Fig 07 Green walls in west side of Qsquare 2
图 08 多媒体塔绿墙
Fig 08 Multimedia tower green walls in Qsquare

Plants: Asparagus setaceus, Pteris ensiformis cv. victoriae, Rhoeo spathacea (Sw.) Stearn, Syngonium podophyllum, Asplenium nidus Linn, Philodendron, Epipremnum pinnatum, Dracaena marginata, etc.

桃园国际机场
TAIWAN TAOYUAN INTERNATIONAL AIRPORT TERMINAL 1, TAOYUAN, TAIWAN

新材料 NEW MATERIALS

桃园机场为台湾重要的对外门户。于南北廊道间，设置多条块状绿带，以利室内空气清新及内部美化（图09-10）。

名　　称：桃园国际机场第一航站绿墙
位　　置：桃园县大园乡航站南路9号
空间型态：公用空间
绿化面积：约160m²
设置目的：长久性墙面绿化
使用植栽：蔓绿绒、鸭趾草、合果芋、山苏、波士顿蕨、黄金葛、千年木等

Taoyuan International Airport is the most important international airport in Taiwan. Along the hallways that run from the south end of the airport to the north end, there are several green walls which add to indoor landscaping and purify the air (Fig 09-10).
Name: Taiwan Taoyuan International Airport Terminal 1 Green Walls
Location: No.9, Hangzhan S. Rd., Dayuan Township, Taoyuan County 337, Taiwan (R.O.C.)
Space type: Public
Area Size: ≈ 160m²
Purpose: Long-term vertical gardening
Plants: Philodendron, Rhoeo spathacea (Sw.) Stearn, Syngonium podophyllum, Asplenium nidus Linn, Nephralepis exaltata cu. Bastaniensis, Epipremnum pinnatum, Dracaena marginata, etc.

图 09 南北通廊绿墙1
Fig 09 Green walls in south-to-north hallways in Taoyuan International Airport Terminal 1

图 10 南北通廊绿墙2
Fig 10 Green walls in south-to-north hallways in Taoyuan International Airport Terminal 2

高雄中油林园厂行政大楼中庭
CHINESE PETROLEUM CORP. ADMINISTRATION BUILDING, LINYUAN, KAOHSIUNG

中油为台湾石化产业的龙头，于室内安置绿墙，其目的在栽种可吸附家具、复印机、建材、油漆等所产生的室内有害物质，具有净化空气及降低二氧化碳、改善室内落尘的效果（图11-13）。

名　　称：高雄中油林园厂行政大楼中庭绿美化
位　　置：高雄市林园区石化二路3号
空间型态：办公空间
绿化面积：约133m²
设置目的：长久性墙面绿化
使用植栽：鸭趾草、合果芋、白鹤芋、山苏、波士顿蕨、黄金葛、紫锦草等

Chinese Petroleum Corp. is the leading company in Taiwan's petroleum industry. Indoor green walls were set in the company's lobby in order to help remove hazardous air-borne substances releasing from furniture, office equipment, building materials and painted walls. The green walls also help reduce dust fall (Fig 11-13).
Name: Chinese Petroleum Corp. Administration Building, Linyuan Lobby Gardening
Location: No.3, Shihua 2nd Rd., Linyuan Dist., Kaohsiung City 832, Taiwan (R.O.C.)
Space type: Office
Area Size: ≈ 133m²
Purpose: Long-term vertical gardening
Plants: Rhoeo spathacea (Sw.) Stearn, Syngonium podophyllum, Spathiphyllum kochii, Asplenium nidus Linn, Nephralepis exaltata cu. Bastaniensis, Epipremnum pinnatum, Setcreasea purpurea Boom, etc.

图 11-13 行政大楼中庭绿墙
Fig 11-13 Green walls in Chinese Petroleum Corp. administration building lobby

上海赢嘉广场
YING-JIA SQUARE, SHANGHAI, CHINA

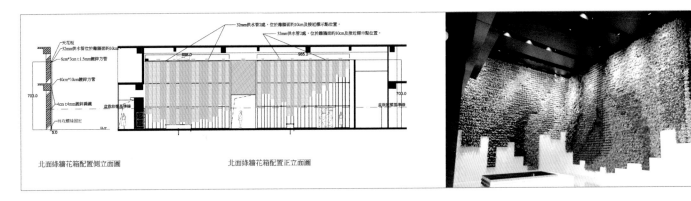

名　　称：上海赢嘉广场绿墙（图14-16）
位　　置：上海市闵行区顾戴路中春路口
空间型态：商业空间
绿化面积：约150m²
设置目的：长久性墙面绿化
使用植栽：鸭脚木、红网纹草、山苏、波士顿蕨、菠
　　　　　萝花、黄金葛、凤尾蕨、竹芋等

Name: Shanghai Ying-Jia Square Green Walls (Fig 14-16)
Location: Cross of Zhongchun Rd. and Gudai Road, Minhang District, Shanghai, China
Space type: Business
Area Size: ≈ 150m²
Purpose: Long-term vertical gardening
Plants: Schefflera heptaphylla, Fittonia verschaffeltii cvs, Asplenium nidus Linn, Nephralepis exaltata cu. Bastaniensis, Erwinia ananas, Epipremnum pinnatum, Pteris ensiformis cv. victoriae, Calathea makoyana, etc.

图 14-16 上海顾戴路赢嘉广场绿墙
Fig 14-16 Green walls in Shanghai Ying-Jia Square

上海佘山月湖会馆
HOMA MOON LAKE HOTEL OF MODERN ART, SHANGHAI, CHINA

名　　称：上海佘山月湖会馆绿墙（图 17-21）
位　　置：上海市松江区佘北路
空间型态：酒店房间绿化
绿化面积：约 140m²
设置目的：长久性墙面绿化
使用植栽：鸭趾草、非洲菊、吊兰、千年木、狼尾蕨、白鹤芋、广伞枫等

Name: Shanghai Homa Moon Lake Hotel of Modern Art Green Walls (Fig 17-21)
Location: Shebei Road, Songjiang District, Shanghai, China
Space type: Business (hotel)
Area Size: ≈ 140m²
Purpose: Long-term vertical gardening

Plants: Rhoeo spathacea (Sw.) Stearn, Gerbera jamesonii, Chlorophytum comosum, Dracaena marginata, Davallia mariesii Moore ex Bak, Spathiphyllum kochii, Heteropanax fragrans (Roxb.) Seem, etc.

-21 上海佘山月湖会馆绿墙
-21 Green walls in Shanghai Homa Moon Lake Hotel of Modern Art

作者简介:

谢宗钦／男／顾问／台湾南海股份有限公司／台湾

Biography:

Tzung-Chin Hsieh / Male / Consultant / Taiwan Nanhai Inc / Taiwan

缘起天开
TianKai Story

"虽由人作，宛自天开"是中国古代造园大师计成在《园冶》中的名句，道出了造园的最高境界。以质量立身，以景效立命的天开集团恪守这份造园理念，抱着"为有限的城市空间营造无限的自然理想"的初衷，励志将公司打造成为"人居环境景观营造"的标杆式企业。天开园林首家公司由现任总裁陈友祥和副总裁谭勇于2003年在重庆联手创立，从此开启了天开时代的奋斗历程。2004年北京天开公司成立，天开集团进一步加快前进步伐，业务量逐年稳步增长，公司知名度也在大幅提升。2012年，天开完成了全国布局和业务升级，分公司覆盖中华各大区，北京、天津、上海、重庆、成都、长沙、哈尔滨等地。源于对景观工程的质量和景效的高要求，天开在行业内获得众多如万科、龙湖、纳帕、中旭、泰禾等高端合作伙伴的信赖，2012年3月，蓝光和骏地产也正式与天开签约结为战略联盟。时至今日，天开已发展成为一家集园林景观设计、园林工程施工、家庭园艺营造、苗木资源供应及石材加工为一体的领军式园林公司。我们的实力、品质和战略布局，能为客户有效降低沟通管理成本，更是全国地产企业所有项目卓越品质的保障。

China has a long history in Gardens and Landscape. "Materpiece of Nature, although Artificial Gardens", as the first and the most acknowledged philosophy about garden building in the world, was initiated from the book Yuanye which was written by Ancient Chinese gardening master Jicheng, and Tiankai was named after it. Tiankai company was established in Chongqing in 2003 by President Chen youxiang and Vice-president Tan yong. We come from nature and back to natue, that's the reason why we love nature gardens. Tiankai hopes to create infinite natural idea for the limited city. Today, Tiankai has become to comprehensive group corporation including garden construction, design, plants maintenance, seedling resources and so on. Tiankai has the second class qualification of garden construction, and B class qualification of landscape design. There are many international and outstanding designers in Tiankai. Tiankai is the best company in China in garden design and construction field. Tiankai has established 10 branches respectively in Beijing, Shanghai, Tianjin, Chongqing, Chengdu, Changsha and Haerbin. Tiankai's contruction business covers all over the country. The partners of Tiankai include Vanke, LongFor, Napa and Blue Ray, which are all very famous enterprises in China. We can effectively save communication and management cost for our clients and keep all the projects in consistent and excellent quality.

www.tkjg.com   tiankai@tkjg.com

成都 蓝光 观岭国际

虽由人作 宛自天开

Masterpiece of Nature
Although Artificial Gardens

天开园林
TianKai Landscape

天开园林咨询：4000-577-775　私家别院咨询：4000-615-006

北京　上海　天津　成都　重庆　长沙　哈尔滨

2012中国优秀园林工程金奖——中国铁建国际城(杭州)示范区园林绿化工程

杭州市园林绿化工程有限公司
HangZhou Landscape Garden Engineering Co., Ltd.

企业资质
城市园林、市政公用工程施工总承包壹级
园林古建筑工程专业承包贰级　城市及道路照明工程专业承包贰级
风景园林工程设计专项乙级　建筑装饰装修工程设计专项乙级
绿化造林设计(施工)乙级　房屋建筑工程施工总承包叁级
土石方工程专业承包叁级　建筑装修装饰工程专业承包叁级

大地融合演绎人居佳境　长育奇葩业绩造品质生活

杭州市园林绿化工程有限公司创建于1992年，现已发展成集园林市政规划设计、工程项目施工、花卉种苗研发生产为一体的企业集团，拥有近十家全资子公司和控股公司，业务范围遍及全国。

杭州园林以创新进取的态势拓展多元化绿色产业，持续提升管理，加快市场拓展步伐，逐步成长为行业的领军企业，位居建设部白皮书大型项目施工能力排名全国第二名，综合排名全国十强企业；具备国家园林一级、市政工程施工总承包一级、风景园林设计专项乙级等十余项资质。

通过ISO9001质量管理、ISO14001环境管理、OHSAS18001职业健康安全管理体系认证。工程施工年产值超十亿元，承建的工程先后获得"鲁班奖"（国家优质工程）、"中国风景园林金奖"、"钱江杯"、"省、市优秀园林绿化工程金奖"等众多荣誉。先后被评为"国家高新技术企业"、"全国城市园林绿化企业50强"、"全国农林水利系统劳动关系和谐企业"、"中国十大创新型绿化观赏苗木企业"、"省级园林业重点龙头企业"等荣誉，连续12年荣膺浙江省"AAA级守合同、重信用"企业。

杭州园林为浙江省桂花种质资源保育与利用公共基础条件平台重要成员，种苗研发基地被授予"中国桂花品种繁育中心"、"杭州市高新技术研发中心"等荣誉称号，组织攻关的杭州市科技进步一等奖"桂花品种基因库建设及优良品种选育推广"荣获2007年度《桂花种质创新及促成栽培关键研究与示范》。

浙江省重大科技项目《桂花种质创新及促成栽培关键研究与示范》。

企业承担了大量的社会责任，履行着十余家行业协会的领导职能，如中国花协桂花分会、中国花协绿化观赏苗木分会、浙江省风景园林学会、浙江省植物学会园林植物分会、浙江省花协绿化苗木分会等。

公司地址：杭州市凯旋路226号浙江省林业厅6F/8F　联系电话：0571-86095666/86431126　传真：0571-86097350　邮编：310020　网站：www.hzyllh.com

盛世绿源
SHENGSHILVYUAN

盛世绿源
彩色城市

为什么说盛世绿源是专业的彩色树供应商？

1、种苗自北美原生区引进，片植、孤植均有极好的造景效果

2、五大彩色树种植基地：总面积2万余亩，苗木50余万株，满足客户的需求

3、不同温度带的适生基地，保障了树木在国内各地区能够健康生长

4、完备的技术措施，保证了苗木的成活率

红枫、红栎、海棠、白蜡、山楂

诚信 责任 卓越

联系我们：0411-83705888
　　　　　　　　83670830
　　　　　　　　83659060

盛世绿源科技有限公司
大连市西岗区北京街126号

SUN&PARTNERS INCORPORATION

PROFESSION CHANGE LIFE

广州山水比德景观设计有限公司

湘江一号实景图

广州珠江新城·珠江北岸 文化创意码头
领航创意 共创未来

广州市天河区珠江新城临江大道685号红专厂F19栋
TELL:020-37039822 37039823 37039825
FAX:020-37039770
E-MAIL:SSBD-S@163.COM

全国招
TEL:020-37039822
FAX:020-37039770

《世界园林》征稿启事
Notes to Worldscape Contributors

1. 本刊是面向国际发行的主题性双语（中英文）期刊。设有作品实录、专题文章、人物／公司专栏、热点评论、构造、工法与材料（含植物）5个主要专栏。与主题相关的国内外优秀作品和文章均可投稿。稿件中所有文字均为中英文对照。所有投稿稿件文字均为 Word 文件。作品类投稿文字中英文均以 1000–2500 字为宜，专题文章投稿文字的中英文均以 2500–4000 字为宜。

2. 来稿书写结构顺序为：文题（20字以内，含英文标题）、作者姓名（中国作者含汉语拼音，外国作者含中文翻译）、文章主体、作者简介（包括姓名、性别、籍贯、最高学历、职称或职务、从事学科或研究方向、现供职单位、所在城市、邮编、电子信箱、联系电话）。作者两人以上的，请注明顺序。

3. 文中涉及的人名、地名、学名、公式、符号应核实无误；外文字母的文种、正斜体、大小写、上下标等应清楚注明；计量单位、符号、号字用法、专业名词术语一律采用相应的国家标准。植物应配上准确的拉丁学名。扫描或计算机绘制的图要求清晰、色彩饱和，尺寸不小于15cm×20cm；线条图一般以 A4 幅面为宜，图片电子文件分辨率不应小于 300dpi（可提供多幅备选）。数码相机、数码单反相机拍摄的照片，要求不少于 1000 万像素（分辨率 3872×2592），优先使用 jpg 格式。附表采用"三线表"，必要时可适当添加辅助线，表格上方写明表序和中英文表名，表序应于内文相应处标明。

4. 作品类稿件应包含项目信息：项目位置／项目面积／委托单位／设计单位／设计师（限景观设计）／完成时间。

5. 介绍作品的图片（有关设计构思、设计过程及建造情况和实景等均可）及专题文章插图均为 jpg 格式。图片请勿直接插在文字文件中，文字稿里插入配图编号，文末列入图题（须含中英对照的图号及简要说明）。图片文件请单独提供，编号与文字文件中图号一致。图题格式为：图 01 xxx/Fig 01xxx。图片数量 15–20 张为宜。可标明排版时对图片大小的建议。

6. 文稿一经录用，即每篇赠送期刊 2 本，抽印本 10 本。作者为 2 人以上，每人每篇赠送期刊 1 本，抽印本 5 本。

7. 投稿邮箱：Worldscape_c@chla.com.cn 联系电话：86-10-88364851

---

1. Worldscape is an international thematic bilingual journal printed in dual Chinese and English. It covers five main columns including Projects, Articles, Masters / Ateliers, Comments, and Construction & Materials (including plants). The editors encourage the authors to contribute projects or articles related to the theme of each issue in both Chinese and English. All submissions should be submitted in Microsoft Word (.doc) format. Chinese articles should be 1000-2500 characters long. English articles should be 2500-4000 words long.

2. All the submitted articles should be organized in the following sequence: title (no more than 20 characters and the English title should be contained); author's name (for Chinese authors, pin yin of the name should be accompanied; for foreign authors, the Chinese translation of the name should be accompanied if applicable); main body; introduction to the author (including name, gender, native place, official academic credentials, position/title, discipline/research orientation, current employer, city of residence, postal code, E-mail, telephone number). For articles written by two or more authors, please list the names in sequence.

3. All persons, places, scientific names, formulas and symbols should be verified. The English submissions should be word-processed and carefully checked. Measuring units, symbols, and terminology should be used in accordance with corresponding national standards. Plants should be accompanied with correct Latin names. Scanned or computer-generated pictures should be sharp and saturated, and the size should be not less than 15cmx20cm. Diagrams and charts should be A4-sized. The resolution of digital images should be not less than 300dpi (authors are encouraged to provide a selection of images for the editors to choose from). The resolution of pictures generated by digital camera and digital SLR camera should be not less than 3872x2592, and .jpg formatted pictures are preferred. Annexed tables should be three-lined, and if necessary, auxiliary lines may be used. All tables should be sequenced and correspond to the text. Chinese-English captions should be contained.

4. All the submitted materials should be accompanied with short project information: site, area, client, design studio (atelier or company name), landscape designers (landscape architects) and completion date (year).

5. All project images (to illustrate the concept, design process, construction and built form) should be .jpg format. The images should be sent separately and not integrated in the text. All images should be numbered, and the numbers should be represented in the main body of the text. At the end of the text, captions and introductions to the images should be attached (Chinese-English bilingual text). The caption should be formatted as Fig 01 xxx. No more than 20 images should be submitted. Suggestions to image typeset may be attached.

6. The author of each accepted article will be sent 2 copies of the journal and 10 copies of the offprint. In the case of articles with 2 or more authors, each author will be sent 1 copy of the journal and 5 offprints.

7. Articles should be submitted to: Worldscape_c@chla.com.cn Tel: 86-10-88364851

青草地园林市政

——营建城市新型绿地　探寻花卉发展模式

浙江青草地园林市政建设发展有限公司是具有国家壹级城市园林绿化企业资质、市政公用工程施工总承包贰级、绿化造林施工资质乙级、绿化造林设计资质乙级、园林古建筑工程专业承包叁级、河湖整治工程专业承包叁级、城市及道路照明工程专业承包叁级、体育场地设施工程专业承包叁级、机电设备安装工程专业承包叁级，集园林绿化设计施工、园林植物科学研究、花卉生产销售、园林信息咨询和鲜花礼仪服务一体的综合型园林市政企业。

湖州潜山公园

海天公园

海天公园假山景观

湖州潜山公园

中华树艺苑

　　公司确定以"质量立业"为发展定位，辨证地处理量的跨越和质的提高的关系。先后承接了海天公园（包括海天高尔夫球场）整体绿化工程、杭州樱花小筑室外景观工程、萧山区风情大道北伸绿化工程、丽水市滨江景观带工程、红谷滩新区总体绿化工程、南昌市象湖公园景观工程、湖州潜山公园景观工程、湖州南浔嘉沁园室外总布景观绿化工程、湖州大剧院景观工程、江苏省太仓港口开发区管理委员会、绿城·桂花园一起景观绿化工程、上海大上海会德丰广场硬景工期、上海新华路一号景观工程、银亿海尚广场、乐清东山公园一期建设工程、上海三甲港江畔御庭别墅绿化工程、温州广汇景园绿化工程等400余项，多项工程获得过"杜鹃花奖"、"百花奖"、"茶花杯"、"最佳人居住环境奖"、"飞英杯"、浙江省优秀园林"金奖"等奖项，绿化市场逐渐向全国拓展。

和泉中菜城一期景观

温州广汇景观

独特的视角
创新的设计

城市规划·景观设计·建筑设计·项目策划

地址：北京市朝阳区北苑路甲13号北辰新纪元大厦2座1705室
电话：010-8492 1962/ 8492 7362　　传真：010-8482 8780
邮箱：marketing@macromind.cn　　网站：www.macromind.cn